Scientific Misconduct Training Workbook

Global Science Education

Series Editor
Professor Ali Eftekhari

Learning about the scientific education systems in the global context is of utmost importance now for two reasons. Firstly, the academic community is now international. It is no longer limited to top universities, as the mobility of staff and students is very common even in remote places. Secondly, education systems need to continually evolve in order to cope with the market demand. Contrary to the past when the pioneering countries were the most innovative ones, now emerging economies are more eager to push the boundaries of innovative education. Here, an overall picture of the whole field is provided. Moreover, the entire collection is indeed an encyclopaedia of science education and can be used as a resource for global education.

Series List:

The Whys of a Scientific Life
John R. Helliwell

Advancing Professional Development through CPE in Public Health
Ira Nurmala and Yashwant Pathak

A Spotlight on the History of Ancient Egyptian Medicine
Ibrahim M. Eltorai

Scientific Misconduct Training Workbook
John G. D'Angelo

The Whats of a Scientific Life
John R. Helliwell

Scientific Misconduct Training Workbook

John G. D'Angelo

CRC Press
Taylor & Francis Group
Boca Raton   London   New York

CRC Press is an imprint of the
Taylor & Francis Group, an **informa** business

CRC Press
Taylor &Francis Group
6000 Broken Sound Parkway NW, Suite 300
Boca Raton, FL 33487-2742

First issued in paperback 2022

ISBN-13: 978-0-367-22345-8 (hbk)
ISBN-13: 978-1-03-233814-9 (pbk)
DOI: 10.1201/9780429274466

**Visit the Taylor &Francis Web site at
http://www.taylorandfrancis.com**

**and the CRC Press Web site at
http://www.crcpress.com**

Contents

Preface

THIS WORKBOOK, ITS COMPANION PowerPoint presentation online, instructor's manual, and website are intended to give participants an understanding and awareness of ethical issues as they relate to research (especially science) and the publishing of research. Upon completion of the program, including submission of the assessments and affirmation statement (outlined online), students will be awarded a certificate of completion.

I suggest that students work in small groups to discuss their responses to the interactive assignments in this workbook. Students would ideally, but not necessarily, come with solo responses to these scenarios already worked through. The groups will then reconvene as a whole class and discuss responses to the assigned cases. At the conclusion of this discussion, Assessment Test #1 would be administered and subsequently discussed and collected. Assessment Test #2 would then be administered, and the Affirmation Statement distributed and collected. All collected materials, along with a processing fee (see website for pricing structure and contact information; https://sites.google.com/site/johngdangeloalfredu/ethics-seminar) will then be sent to John D'Angelo, Ph.D. (see website for contact information; https://sites.google.com/site/johngdangeloalfredu/ethics-seminar) for processing and final distribution of the certificates.

Despite the abhorrent nature of interpersonal workplace conflicts, issues such as discrimination and all forms of harassment and assault will not be rigorously covered here, though cases do cover collaborations. These issues go well beyond scientific/research ethics and are often a violation of state and/or federal law. Your employer or sponsor is likely to provide training in company or institutional policies regarding these issues; such training is mandatory for institutions receiving any manner of federal monies, as of this writing.

HOW TO USE THIS BOOK

Although there is minimal space to do so, I strongly recommend that you take notes in this book, particularly when you are responding to questions about individual cases. They will help you to better remember what you were thinking for each case. This will help you to grow into a more thoughtfully ethical scientist. If you do not interact with the book in this way, it is unlikely to have the desired impact.

In the scenarios in this book, I have tried very hard to only use gender-neutral pronouns. This is to try to be as inclusive as possible. I have also worked very hard to word some of the pondering points in a way to stimulate discussion and, in some cases, even tried to play devil's advocate and take up a potentially unpopular or controversial stance. This is deliberate and is not meant to insinuate I am taking a particular side.

I believe this book would work best in a setting (in-person or online) in which you are able to interact with other people in discussing the cases. You should record in some way your thoughts on each case and then also take notes during the discussion with your peers and/or instructor. For each of the cases, you should record your thoughts and then also record how those thoughts are reinforced or changed by any and all ensuing discussions. Nothing less will allow this work to have the intended impact.

ONLINE CONTENT

PowerPoint

A PowerPoint presentation created by the author for instructors to use as is or in a customized form in the delivery of training is available.

Instructor's Guide

The workbook, along with some points for guided discussion, is available to instructors to further explain the hypothetical cases. This is omitted from the workbook to avoid it being like a text.

Assessment Docs

A primary assessment exercise that probes the learners' ability to identify definitions or descriptions of scientific/research misconduct is available for instructors seeking to evaluate learning in the workbook. A deeper assessment assignment that is designed to be given as a before and after set to probe growth in the learners' thinking and reactions to especially grey cases. This deeper assessment is not mandatory to receive the certificate. Results from it, however, would potentially be part of a publication describing the seminar.

Instructor's Guide For Assessment Docs

An answer key for the primary assessment and a guide on how to use the deeper assessment is provided.

Case Studies

Files that provide an overview of a recent or famous case of (at least alleged or potential) scientific or research misconduct along with discussion-inspiring questions and an associated instructor's version are available.

Videos

These videos are housed by YouTube at the author's channel, Dr. D'Angelo's Scientific Ethics (https://www.youtube.com/channel/UCZYKQZQQAoaU--wLmVVIyyg). They are short lectures that discuss issues of science, scientific/research misconduct, contemporary science issues in the public, famous cases of misconduct, priority disputes, and more.

CHAPTER 1

What Is Science and How Is Science Done?

S CIENCE CAN BE … capricious, if you'll permit me to anthropomorphize. Sometimes, your best laid out experiments will epically fail. Other times, they work so well, you need to pinch yourself to verify you aren't dreaming. Still other times, you obtain a result that is actually far more interesting than any result you expected. I'd argue the latter is where some of the best science happens, in fact; but it sometimes is the hardest to achieve since it requires a wide-open mind.

To try to cultivate this open mind in my students, I tell them that the only failed experiment is the one you learn nothing from. For example, in synthetic chemistry, it is critical to keep in mind that if a reaction fails to give any product, this doesn't mean that you didn't get a result; you just didn't get the result you expected and wanted. Analogies to all forms of scientific and/or engineering research can easily be drawn. What is essential in these cases across all fields is that there is a *reason* for the failure and identifying that reason gives you something different to try next time. Such iterative experimentation, I'd argue, is the very essence of science.

Incremental improvements in materials, instruments, methods, or medicines fit into this scenario as well, even when fantastic success is initially enjoyed. The incandescent light bulb endured for a long, long time. As materials have improved and energy consumption has become a global concern, new light sources have been developed based on newer and more efficient technology. The miniaturization of materials has resulted in my smartphone having a larger capacity than that of my first four or five desktop computers combined. Penicillin has saved countless lives, yet drug resistance is a problem of worldwide importance; science continues to deliver effective antibiotics.

One of the key aspects of science that it is largely self-corrective. This can be said about how breaches of the above-mentioned norms of behavior are discovered and also of genuine mistakes made being corrected. The history of science abounds with many a failed theory. Some of these theories (like that of phlogiston) endured longer than others. By no means do errant theories constitute a breach of behavioral norms. Rather, they (or perhaps their eventual fall) represent science working at its very best. As breakthroughs in observational techniques or fabrication processes improve, so does our data and, inevitably, the accuracy of our conclusions. As such, we must talk about what science is and how it is done before what constitutes as misconduct can be considered and understood.

Science does all of this through a long, arduous, and usually expensive process of trial and error. One of the most difficult things to get used to, especially for a younger scientist, is that it often takes many tries to arrive at success. Frustration and losing patience with this are likely a contributing factor to at least some instances of modes of misconduct, but this is only a suspicion I have; I have never performed a study to psychoanalyze why people perpetrate misconduct.

Now, let's consider some hypothetical cases of science and try to identify where it's working well.

HYPOTHETICAL CASE STUDY: DISCOVERY AND DEVELOPMENT OF DRUGS

A study where the researchers are trying to improve the efficacy of an anticancer drug starts with a lead compound identified using computer-based modeling software. Testing is done *in vitro* against multiple cancer cell lines and then the molecule is modified, changing one of three chemical groups one at a time. To complete this initial screening takes a year and a half. In the next round, each of the best five substituents is combined in all of their permutations (125 total). They then move the best five analogs forward and begin animal testing. Finally, seven years after starting the study, the analog with the best combination of activity, safety, and ease of synthesis is moved into phase I (in healthy people) clinical trials in humans.

Pondering Points

- Why couldn't the process go faster?

- What could have been done to speed along the process?

- What are some potential pitfalls to going faster?

- Why would the ease of synthesis be something that would determine which analog moves forward?

- Why isn't it unethical to include healthy volunteers?

Wrapping Up

Drug discovery is typically a long and arduous process. Rarely is it completed faster than several years, although modern FDA regulations (in the U.S.) allow for fast-track studies under well-defined circumstances. An iterative process that combines most of the active structural units is common. Additional studies that measure how drugs are metabolized and how long they stay in the body are also routinely done. Successful drugs must balance all such properties. For example, a drug that has the most outstanding activity profile in an *in vitro* study may in fact fail *in vivo* if it is eliminated from the body too quickly. This is not always possible to know after *in vitro* studies. Alternatively, synthetic processes to prepare the drug may prove so costly to do on a large scale that the drug would become too expensive. Many factors, therefore, contribute to the success or failure of a drug candidate. A full discussion of these factors is beyond the scope of this workbook.

HYPOTHETICAL CASE STUDY: WOMEN IN DRUG TESTING

Consider the following two scenarios:

During the clinical trials of a powerful antibiotic, a lower number of women were enrolled. After several years on the market, it comes under scrutiny after routine post-market safety evaluations reveal that women who have taken this drug are 1% more likely to develop ovarian cancer than women who have not taken this drug. In consultation with the FDA, the company that produced the drug modifies the warning label but the drug stays on the market.

During clinical trials, an equal number of men and women are enrolled. The study proceeds and finds no serious negative side effects and also shows outstanding efficacy at treating migraine headaches. Years after the study, and after the drug has been deployed for use, a disturbing trend is found: A significant portion of the women in the study are suffering miscarriages. Each of the women is diagnosed with the same, rarely seen uterine abnormality. The possibility of women with the abnormality being part of the study is ruled out; the compound being tested is believed to have caused the abnormality.

Pondering Points

- Each of these hypothetical situations is very bad. Is one of them worse?

- Assume instead that the second case caused terrible birth defects, rather than miscarriages. That is, the woman is fine, but the baby has birth defects that ensure a painful and short life and that these defects are not detectable before birth, at least not with current technology. Does your opinion on which one is the worse change?

Wrapping Up

The disproportionately lower number of women in clinical trials has recently begun to receive attention.* It shouldn't be a surprise that the biology of men and women is at least slightly different, if not for any other reason apart from the fact that women have organs men do not and vice versa. Additionally, women can bear children while men cannot. While men produce sperm from the onset of puberty, women do not constantly regenerate eggs—meaning anything that damages but doesn't destroy an egg can go on to create a fetus with abnormalities, but damaged sperm is likely to be replaced with healthy sperm, provided it wasn't the germline cells producing the sperm that were damaged. Furthermore, there are truly ethical questions regarding the administration of an untested compound to a pregnant woman as its effects on the unborn baby are unknown. This brings about a complication in that a woman who becomes pregnant during testing may need to be removed or may remove herself from the study. Otherwise, it may only be found out the hard way that such issues are side effects. Of course, if women are left out of the trials, these side effects may only be found out later, after more women (or their offspring) are suffering from it. Ordinarily, these sorts of side effects are seen during the animal testing phase. As we wean ourselves off such testing, however, this problem may get worse. In short, testing on women and even administering treatments to women could impact at least more than one life. Thus, it is often

---

* Liu KA, DiPietro Mager NA. Women's involvement in clinical trials: historical perspective and future implications. *Pharmacy Practice* 2016 Jan–Mar;14(1):708.

wise to apply extra precautions. Of course, if these side effects are going to happen, they must be found either before or after a drug hits the market.

HYPOTHETICAL CASE STUDY: CHILDREN IN DRUG TESTING

Imagine the following scenarios:

A new oral allergy medicine is available at a lower dose in solution for children. During testing in adults, the drug had no serious side effects. After being in use for several years, children taking the medicine are increasingly diagnosed with asthma. The development of asthma is not observed in adults and generally abates a few days after the children stop taking it.

A study is being done to determine if funny stories are more memorable than scary stories using an IRB-approved protocol. A wide range of ages of subjects are recruited and informed consent is obtained in all cases. The parents of the children in the scary story group, who were clearly informed that their children would be subjected to scary stories, complain to the study's designers that their children are now suffering from night terrors where they previously had not. The study's designers put into motion their protocols to provide counselors to assist these children.

A study that is testing a new anti-leukemia drug involves many children since this form of leukemia is known to affect kids and adults. No serious side effects are observed at the time of testing. However, as the children grow older, a significant percentage of them begin to start puberty much younger, around age seven or eight.

Pondering Points

- Are any of these situations more acceptable than others?

- Are any of these situations less acceptable than others?

- All of these situations are not very life-threatening. Where is your threshold for acceptable risks regarding children, keeping in mind that they may not understand the study as completely as adults can and also keeping in mind that, in the real world, kids do get sick, sometimes very sick, and are also subjected to frightening things.

Wrapping Up

Suffering children is something that pulls at the heartstrings of the vast majority of civilized society. Children are one of several protected groups of people during human subjects research. This is for at least a few reasons. First, most, if not all children are not able to fully understand the risks and so they are of limited ability to make an informed decision. A legal guardian or parent makes that decision for them and in cases where there is a side effect, the *child*, no matter how much said guardian feels guilty, suffers much more. Also, since children are not done developing, especially sexually, such development may be impacted by medication. Of course, if these side effects are going to happen, they must be found, either before or after a drug hits the market preferably before. Before is by far the lesser of two evils, since it will impact fewer children. Side effects that impact a child's development would never be observed in a fully developed adult. Children also have at least slightly different metabolisms, which will make the body weight to dose ratio potentially different than it is with adults. Even with non-pharmaceutical testing, children react to and cope with stimuli in a different way than adults. These differences make the inclusion of children in such studies necessary, but only with rigorous communication to parents or guardians and strong, rapidly accessed safeguards as well. Remember though, that in the case of pharmaceutical testing, some measure of safety will have been observed, at least in animals, before it is tested on people, at least for now.

HYPOTHETICAL CASE STUDY: BRIBING DOCS BY PHARMA

A pharmaceutical representative meets with a doctor's office about a new strep throat test that doesn't require swabbing the patient's throat. When dropping off a free box of recently FDA-approved sample tests, the representative brings several trays of hot food from a nearby fancy restaurant. The rep shares with the nurses, physician's assistants, and doctors in the office the peer-reviewed paper describing the effectiveness of the new tests along with documentation shared with the FDA regarding the safety and efficacy, before giving a short presentation and demonstration on how to use them.

Pondering Points

- Did the representative do anything untoward?

- Assume that, instead of a fancy restaurant, the food was simply take-out from a local deli or fast food. Does your impression of what the representative did change?

Wrapping Up

This sort of behavior, although it may raise eyebrows, is not inappropriate. Reps of all kinds are permitted to take perspective users of their products to lunches, dinners and a variety of other "perks." What is not permitted, on the other hand, is to offer some measure of kickback on a per prescription or use basis. Can one make the argument that it is impossible to be impartial after receiving some sort of outstanding meal or fancy new electronic device? I suppose you can make that argument. However, if all of the companies are doing it for their products, isn't the playing field level? Also, these business tactics are fundamentally different than some sort of bonus, or other benefit, afforded each time a product is used.

HYPOTHETICAL CASE STUDY: UNFAIR PRICING

A struggling pharmaceutical company hires a new chief financial officer (C.F.O.). As part of their efforts to place the company on stronger financial grounds, the new C.F.O. looks carefully at the balance sheets and notices that one drug is perennially deep in the red since there are so few patients who use this drug; a medicine for a rare disease. They resolve that the price must be increased stepwise for the next few years to get the sheets balanced for this product, ultimately increasing the price by 750% over five years.

Pondering Points

- Is it the responsibility of the C.F.O. to make sure the company is healthy (financially) or the public is healthy?

- An annual price increase rate of 150% means that the price will more than double every year for five years. Does this stepwise increase make it OK since it gives the consumer and insurance companies time to compensate and plan? (If the starting cost was $30, the final cost would be $255.)

- Does the fact that it is only a relatively small number of patients that are paying the price for this increase matter?

Wrapping Up

If the pharmaceutical industry as a whole goes bankrupt, nobody gets any medicine. Now, if such an extreme financial calamity were to happen, the odds are good that there is a level of chaos occurring worldwide that may make a lack of medicines seem small. Nevertheless, companies must be allowed to do what is necessary in order to stay solvent. Let us not conflate humanitarianism, or benevolence, with what is right. Surely, we would all like to think that multibillion-dollar companies won't mind spending millions to help people. Expecting, or stronger yet, mandating that they do so is a line that perhaps we'd be better not to cross. If other benefactors want to support such medicines being more widely available, however, mechanisms should be developed to make this possible and easy. Governmental (local, state, or federal) entities could also step in as a matter of protecting public health. The bottom line is that the C.E./F.O. of a pharmaceutical company is not always the pariah they are cast to be.

HYPOTHETICAL CASE STUDY: DTC ADVERTISING

A pharmaceutical company has developed a new treatment to treat baldness. In a commercial promoting this treatment, they show an initially bald man portrayed as being lonely and sad. After using the treatment and growing hair, the man is shown as being happy, playing sports and walking arm in arm with an attractive woman while holding hands with a small child. The narrating voiceover never mentions the man; it talks extensively about how fast patients should see results and lists common side effects, before the video concludes with the attractive woman lovingly running her hands through the man's head of hair.

Pondering Points

- What does the commercial assume and insinuate about baldness and a person's happiness?

- Is the commercial informative regarding important aspects of the drug?

- What does the commercial assume is the sexuality of the actors?

Wrapping Up

Direct-to-consumer-advertising is not permitted in all but a few countries—the U.S. is one of the few that permits it. A debate on whether or not it should be allowed is beyond the scope of this work, but it is certainly an interesting and engaging one. The particular case here contains an additional, more subtle trait: disease mongering. In this scenario, baldness is deliberately presented as a disease. While some people may truly feel psychological harm from being bald, compared to something like HIV or brain cancer, it is borderline laughable to consider baldness a disease. Anyone who consumes any form of media in the U.S. (television, internet, radio, etc.) has heard or seen countless commercials for a number of different afflictions. Some of them are far from life-threatening; rather they impact the quality of the afflicted individual's life. Together, these issues do affect millions upon millions of people and so represent a major cash opportunity for the pharmaceutical industry. Also, generally speaking, commercials like this one assume heterosexuality; it is difficult to make any firm conclusions as to why. It could, for example, be due to the fact that generating a commercial with same-sex couples is likely to (unfortunately still) generate a fair amount of negative press. Alternatively, among other possibilities, it may simply be to their being more heterosexuals than self-identifying non-heterosexuals. That is to say, the company may be playing to its perceived audience.

HYPOTHETICAL CASE STUDY: DRUGS APPROVED DESPITE A LIKELIHOOD TO BE ABUSED

A new drug to treat a disease with few viable treatment options is in the final stages of clinical trials. It was found during clinical trials that a side effect is that it produces an effect like that of amphetamines. There is an anti-drug abuse group that is lobbying against the approval of this drug on the basis that it will encourage abuse. The pharmaceutical company that has made the drug has incorporated multiple abuse-deterrent technologies that are standard tactics in the industry.

Pondering Points

- What should the FDA do?

- Should the opposition to a drug's likelihood to be abused directly relate to the dangers associated with overdose? That is, since opioids are far more dangerous, should opposition to them be higher than to less dangerous drugs?

- Is it right to deny people who would use the drug appropriately because there are people who wouldn't?

Wrapping Up

The opioid crisis is currently at the forefront of national attention in the U.S. For certain, there are major societal issues that we need to collectively overcome. The (unfortunate) fact of the matter remains, however, that in terms of pain management, opioids are *extremely* effective, despite their various dangers. Increasingly, these dangers are a threat to the approval of a new drug by the FDA, though some are getting approved.* It is certainly fair to be in support of this threat and in opposition to this threat. On the opposing hand, these treatments are addictive and even people who initially use them appropriately may fall prey to this and develop a dependence or addiction, or overdose and die. On the supporting hand, for some individuals, there may be no other options that are capable of alleviating pain and/or discomfort and that is a true affront to their quality of life. One can easily ask the question "why should such people be forced to live in pain?" When you consider that drug makers have technology at their disposal to add abuse-deterrent properties to the sample, such an argument is strengthened. But, abuse-deterrent is not abuse-proof. Another concern cited is that doctors may inappropriately prescribe such a medicine. This is not the fault of the pharmaceutical industry, nor is it the fault of any patients. Such behavior is on the doctors; it is they who must amend their ways, rather than keep potentially life-changing treatments from being available. Nevertheless, this debate will rage on. Hopefully, the outcome will be the discovery or invention of a new class of pain killers—one that does not have the dangers associated with opioids (though even Tylenol and Aspirin have their challenges) but does have the effectiveness associated with opioids.

HYPOTHETICAL CASE: ORPHAN DRUG STATUS

A pharmaceutical company is developing a new drug that treats a disease with a high mortality rate that one in 300,000 people has. As the drug proceeds through the full gamut of clinical trials, the FDA accepts the company's request for tax credits for the clinical

---

* https://www.npr.org/sections/health-shots/2018/11/02/663395669/despite-warnings-fda-approves-potent-new-opioid-painkiller (last checked May 5, 2019).

testing and waives the prescription drug user fee. After the clinical trials are completed, the drug is approved for use.

Pondering Points

- Does something as wealthy as a pharmaceutical company deserve a tax break or other kinds of credits for such a drug?

- Considering that it can cost billions of dollars to develop a drug and bring it to the market, does a medicine that will eventually be used by only one in 300,000 people stand a chance at recouping the costs?

Wrapping Up

This is a scenario that is covered by the Orphan Drug Act in the U.S. One important caveat is that the drug may not also be used for other, less rare treatments. That is to say, the hypothetical drug mentioned in this case may not also be used to treat a disease that one in ten people have. Certainly, such a use may be discovered after the fact. In those cases, any measure of support received by the company under the terms in the Orphan Drug Act would end upon the use of the treatment for those purposes. Importantly, the potential drug must still be subject to all of the rigorous testing of any other drug. This can get cloudy, however, since doctors can and do prescribe medicines for things other than what they were originally intended to treat. The company, however, can only advertise the use for diseases that the medicine is approved for. Additional approval can be awarded, but testing demonstrating its efficacy in this new setting must be done and submitted to the FDA

HYPOTHETICAL CASE: ACCELERATED APPROVAL

A new anticancer drug is entering into clinical trials for pancreatic cancer. In trials, it is shown to radically shrink the size of pancreatic tumors and be remarkably safe. The FDA grants the drug approval and the company developing the drug continues to monitor the efficacy of the drug after the fact to evaluate if the drug also prolongs the patient's life. It is found during these phase IV confirmatory trials that if a patient stays on the medication, the tumor size remains small, and no patients taking the medication succumb to the disease, while individuals whose treatment was stopped show tumor growth resumed.

Pondering Points

- The drug was approved for use before the study was finished. What problems could there potentially be with this?

- Tumor shrinkage logically correlates with not succumbing to the cancer; does using its observation as a measure of drug success make sense?

Wrapping Up

The program that allowed the FDA to approve this drug for use originates in 1992 and is called Accelerated Approval. Under this program, a drug evaluation can be based on what is called a surrogate endpoint. A tumor shrinking is one example of such a surrogate endpoint. However, a tumor shrinking does not necessarily equate to a prolongation of life, even if it logically *may*. If it does not prevent the tumor from metastasizing, it may have little to no impact on prolonging a patient's life. Another surrogate endpoint that is commonly used is the changing of one's lipid profile (cholesterol levels) as an indication that it will have positive benefits of staving off cardiovascular disease. This is because high cholesterol levels are strongly associated with heart disease, and also often stroke. Since many people who take such medications are doing so for years, it is unreasonable to keep a drug candidate in clinical trials for nearly a decade just to prove it does what it is intended to do. Enter the surrogate endpoint. By reaching a clinical, measurable endpoint associated with a positive outcome, the outcome can be inferred. Using the phase IV confirmatory trials, the actual effectiveness of achieving that outcome is then measured. In cases where the drug fails to achieve the outcome in a statistically relevant way, approval is withdrawn.

HYPOTHETICAL CASE: DRUGS PRESCRIBED FOR DIFFERENT USE

A new medicine has been approved by the FDA for the treatment of cancer. A physician, who holds an M.D. and a Ph.D. in chemistry recognizes the compound as something they saw years ago presented at a conference as having pain-relieving activity. The doctor reviews the literature to verify they are remembering correctly. The pharmaceutical company developing the drug subsequently discovered outstanding anti-cancer properties and pursued its approval for such use, rather than pain-killing use. A patient of theirs is in chronic pain due to an injury suffered while in the armed forces. All non-opioid painkillers have failed to bring this patient relief. The doctor prescribes the patient this drug as a final effort to alleviate the patient's pain in a safe way.

Pondering Points

- Should the doctor prescribe this medication for this use?

- Should the pharmaceutical company advertise this property of the drug?

Wrapping Up

Although it may seem very strange, once a medication is approved for use, it can be used for anything that prescribers (a medical professional with prescription-writing privileges) decides, even if it is not the use that the medication is approved for. The pharmaceutical company, however, cannot advertise any uses except that which the drug is approved for. It is in fact not uncommon for a drug to have more than one property that is biologically beneficial. Ordinarily, a doctor would not wantonly prescribe something though in the absence of some manner of supporting data and then just hope for the best. Sometimes, prior experience or familiarity with a drug or a similar drug may inspire the doctor to make an educated guess that the drug may work for other indications as well. Other times, they may hear something at a conference or other professional meeting. Whatever the case, a health professional who intends to keep their license would only pursue such a treatment with very good evidence that it will work.

HYPOTHETICAL CASE: MINING LITERATURE FOR PREVIOUS WORK

A psychologist performs a research experiment that has been approved by their institution's Institutional Review Board (IRB) investigating the relationship between parents' level of school-ing and their biological children's IQ and comparing the results to adopted children. After the study is concluded and the results have been analyzed, the researcher finds during their literature review prior to writing a report of the research for publication that the exact study was reported by other researchers at a less elite institution a few years prior. The review allows the researcher to see that there is a logical next step that can be performed; how-ever, their funding was exhausted by the study. They abandon the project and assume the other researchers have also seen this next step and are carrying it forward.

Pondering Points

- What should the researcher have done differently?

- Did the researcher overreact in abandoning the project?

- Should the results obtained be published, even though or perhaps specifically because it confirmed the results of the less elite institution?

HYPOTHETICAL CASE: MINING LITERATURE FOR PREVIOUS WORK AND GETTING DIFFERENT RESULTS

A young psychologist performs a research experiment approved by their institution's IRB investigating the relation-ship between parents' level of schooling and their biological children's IQ and comparing it to adopted children. After the study is concluded and the results are analyzed, the researcher finds that, during their literature review prior to writing a report of the research for publication, the exact study was reported by another researcher. This researcher is a Nobel Laureate. The results that the Nobel Laureate received are opposite those being observed in the present study. The young researcher abandons the project assuming the more established researcher was right and that to get different results indicates a flaw in their own research design.

Pondering Points

- What should the researcher have done differently?
- Did the researcher overreact in abandoning the project?
- Should the results obtained be published, even though or perhaps specifically because it disagrees with the results of a Nobel Laureate?

Wrapping Up

Performing research should always follow the scientific method. No matter how you define or label each individual step, one of the steps *must* be to perform some modicum (at least) of literature research. If the work has already been reported, in the absence of some measure of expected improvement or a significantly novel method, it is not appropriate to repeat the study with the intent to publish. Generally, verifying someone else's results is not considered publishable. This type of literature survey should be done before any lab work is done in order to avoid a situation where you have done nothing more than verify someone else's work. In a case like our second story, however, where carelessness leads to an experiment being re-done, if different results are obtained, I'm tempted to argue that you are intellectually obligated to find out why.

HYPOTHETICAL CASE: DATA MANAGEMENT

Consider the following pair of cases.

The Set Up

A researcher is investigating the efficacy of a chemical's ability to reduce the motility of cockroaches, repeating each experiment three times. Using a protocol that employs two stopwatches (one to measure rest time, the other to measure mobile time), the researcher attempts to quantify their observations.

Case One

They arbitrarily decide that outliers beyond a predetermined limit in each triplet will be removed from the results. This leads to removal of four data points in total. The data points eliminated all turned out to be on extremes and canceled each other out, anyway.

Case Two

They apply standard statistical analysis tests and determine that two data points need to be removed as outliers.

Pondering Points

- Does it make it OK in case one that outliers on both extremes were removed, effectively canceling each other out in the data?

- Which researcher, if either, acted inappropriately?

Wrapping Up

It is not appropriate to arbitrarily choose which, if any, data points to remove. There are tests (among them the Q-test, Student's t-test, and others) that can be used to determine whether a data-point can be eliminated from consideration. Assuming that the extremes can simply be eliminated and that eliminating both extremes "balance out" the data is flatly wrong. Also, these tests are more typically applied to the repeated trials of individual data points. Rarely will any data point be due to only one trial, particularly when instrumentation is used to gather quantitative data. Of course, some measure of common sense can be applied in that if an instrument has visibly malfunctioned (for example, if an autosampler was misloaded or malfunctioned and drew sample from the blank, rather than the intended sample), such a datapoint can be discarded since it was not actually an analysis of the material being studied.

HYPOTHETICAL CASE: CREATING VS. FINDING A CORRELATION

Consider the following pair of cases.

The Set Up

A researcher analyzes data reported by all the baseball teams in Major League Baseball that shows the timing of all concession sales by inning. After pouring over the data for every game played by every team, they find no obvious correlations with the performance of either team, except for one team that shows that higher concession receipts correlate with home team losses. This stadium is renowned for having excellent food.

Case One

They write up their study and insinuate that the goal of their study was to see if the quality of the food impacted home team performance by way of less fan enthusiasm and energy. This is alleged to be because fans are spending too much time getting or eating food, rather than cheering.

Case Two

They explain that they were curious about concession sales and home team success and found nothing except in a case where the food was outstanding, and the team was perennially bad. They found this curious but were unable to propose a data-based reason for this.

Pondering Points

- Both researchers used the same data. Did one of them *misuse* the data?

Wrapping Up

First and foremost, it must be understood that correlation does not necessarily mean causation. There are many things that are unquestionably unrelated that also correlate with one another. A number of humorous ones can be found online.* If taken as true, among the more humorous (to me) examples are: Number of people who drowned by falling into a pool correlates with films Nicolas Cage appeared in; and, Divorce rate in Maine correlates with per capita consumption of margarine. No intelligent person would ever insinuate that either of these may have a cause-and-effect relationship. One must be careful to not claim there is such a relationship when there isn't one; and, it is unlikely that there is one if a mechanism by which the relationship may work is still a mystery; until then, there is only correlation, and even that may be a stretch and coincidence may be more accurate. Furthermore, one must be very careful when simply collecting a large amount of data to analyze and mine for relationships later. This is not inherently inappropriate, per se, but insinuating that such correlations were expected from the beginning certainly is. Collecting large volumes of data, searching for correlations, and then investigating those correlations further, particularly investigating if similar correlations are observed in other data sets, is valid. In such a case, the narrative of *potential* causation can then be more logically built.

HYPOTHETICAL CASE: SERENDIPITY IN SCIENCE

A plant research team is trying to invent a new weed killer. During their investigation, they find that although it is moderately effective at killing weeds, the portions of the field that the test mixture

---

* http://www.tylervigen.com/spurious-correlations (accessed April 11, 2019).

is applied to is overrun with mosquitos. The population of mosquitos exponentially drops as distance from the source increases. The team switches gears and begins researching the mixture as a bait to lure mosquitos into a trap, eventually filing for and receiving patent protection for the use of the mixture this way.

Pondering Points

- Did the researchers do anything inappropriate?

- Does serendipity need to be acknowledged?

- Is serendipity unethical?

Wrapping Up

Sometimes it is better to be lucky than to be good. On the other hand, it can be argued that only the good will see the higher value of serendipity. Ignoring such a chicken vs. egg argument, there is nothing unethical in realizing that the result you are obtaining is more interesting than the result you were expecting. If anything, this may represent an even better application of the scientific method, since it requires that you think outside the box and beyond your (perhaps biased) expectations. It demonstrates an ability to revise your hypotheses or theories based upon the results you are obtaining and formulating new hypotheses or theories that better fit the data. Maintaining an open mind in research is essential to the scientific method.

HYPOTHETICAL CASE: EXPERIMENTAL DESIGN

A researcher is trying to demonstrate that a pheromone-impregnated fishing lure is more attractive to the fish than a live fish. To test their theory, they place 20 largemouth bass in a swimming pool repurposed as a research tank. They then place the 100 lures they created on a system of flotation devices with motion sensors that register every time the bass strike. There are also 10 live bait fish in the pool. The researchers observe, video record, and visually count the number of fishes that are pursued by the bass. The number of interactions with lures is far greater, and the researchers conclude that their lures work. The use of animals for this research is approved and occurs with appropriate oversight from the relevant IRB.

Pondering Points

- What did the researchers do wrong?

- Is there an experimental design that would be better?

Wrapping Up

When designing an experiment such as this, it is critically important that you design the experiment to determine *if* something will work (or happen, etc.); not *that* it will happen. Certainly, this hypothetical experiment is deliberately designed to be overwhelmingly exaggerated in favor of the lure, but the matter of how to design such experiments is certainly a gray area. What would work best in this case is to try a number of different conditions that simulate in a methodical way different relative amounts of live fish vs. lures. Under such conditions, it can be determined at what point each is preferred over the other. In fact, such a design may (in a best-case scenario for this research project) further argue in favor of the lures if even when there are significantly more live fish than lures, the lures are still preferred. Also, peer review in any journal of even marginal quality would likely reject the research as initially described but that is besides the point.

HYPOTHETICAL CASE: ORDER FROM CHAOS

A researcher in a synthetic organic chemistry lab is trying to invent a new chemical reaction using reagents they are familiar with handling. In the first attempt, they mix unmeasured quantities of four different compounds, anticipating the combination of three of them using the fourth as an activating reagent of sorts. Upon analysis, a very small amount of an interesting but unexpected compound is observed. This product, if really a product of the chemical reaction, could only come about by a novel reaction, as the researcher found during their literature review. After carefully considering how this product may have formed, the researcher tries again using specific and measured quantities of the reagents—the yield improves. After changing one of the original reagents and replacing it with a similar reagent with slightly less reactivity, the yield improves more. Finally, the researcher adds a small amount of a catalyst and the reaction proceeds to form this product in very high yield and purity.

Pondering Points

- What could this researcher have done differently? Ask yourself if you're just mad that they got so lucky doing this a way that at least *seems* decidedly unscientific.

- What were some of the potential dangers to the first attempt?

- Could there have been a more logical way to make this discovery?

- What if it is later found that an excess of one of the reagents was used in the initial attempt and that this excess is necessary for the reaction.

Wrapping Up

Although at a first glance, this may appear to be against the scientific method in that it skips any manner of research question and true preliminary research, I would argue that this sort of approach explores a far more general initial question. Perhaps something like "I wonder what would happen if I mix all this together?" could be the question asked here. I concede that the initial, debatably haphazard, approach "feels unscientific." If there was no follow-up, I would in fact agree with such an evaluation. In our story, however, the researcher identifies something interesting and then begins to optimize their results. Here, the scientific method is working very well, and the question clearly becomes "How can I enhance the production of this interesting product?" Although potentially questionable in its beginnings, this study was scientifically carried out. This is not to say that it is *always* appropriate to start a study this way… one can easily conjure examples where it is not. It is important to always keep in mind, however, that the only failed experiment (neglecting the possibility of an accident or other danger) is the one you learn nothing from.

HYPOTHETICAL CASE: COMBINATORIAL CHEMISTRY

A researcher is trying to discover a new anti-malarial compound. In an effort to make a large number of derivatives at once, they combine two classes of compounds on a 10×10 plate. In so doing, they make 87 compounds (13 reactions failed to yield product). To prepare the compounds takes less than an hour of work and the process is done before the worker's lunch break. After lunch they set up the automated purification system and by the time they return in the morning all 87 compounds are pure and sent to the biological team for testing in assays.

Pondering Points

- Is the worker irresponsible for not optimizing the 13 reactions that failed?

- Does this type of work feel like cheating to you?

- Results of the assay are deliberately left out from this story. How does your impression of the work change for each possible outcome, great activity, or terrible activity against malaria?

Wrapping Up

This method of synthesizing a large number of compounds at once is called combinatorial chemistry. Although it may seem like "cheating" or "not earning your product," this is a protocol that is used to greatly increase the number of compounds that can be made in any given period of time. Part of why this is particularly effective in the realm of drug discovery, despite the typically small quantities of product obtained, is that *in vitro* testing of drug candidates can be completed with milligrams of material or less. Thus, this process can allow a researcher to generate many molecules in a short period of time and then (provided they are sufficiently pure) have them tested in a biological assay. This increases productivity and speeds up the research process allowing for more discoveries to be made in less time. Highly active compounds discovered *in vitro* (not in living things) assays can then be further developed or explored.

HYPOTHETICAL CASE: BEGINNING RESEARCHERS

A young researcher is conducting a study that has been approved by the appropriate IRB committees. In the study, they are proposing to use nets to sample birds feeding on a type of berry bush believed to be negatively affected by climate change. The nets, which are cargo nets they borrowed from their uncle's moving truck company have a large mesh size, approximately 3" around the edges of a square. Over the course of two weeks, they only capture larger birds such as crows and raptors. They write a term paper describing the study with the conclusion "Based upon the lack of smaller species being caught in the net, we conclude that climate change has already acted to displace smaller species of birds."

Pondering Points

- What did the researchers do wrong? Was it experimental design, interpretation, reporting of the results, or some combination of all of these?

Wrapping Up

The researchers in this case did not design their experiment very well, though, given their level of (in)experience, perhaps they cannot be blamed for this. Also, this is very different from designing an experiment to only have the larger bird outcome. Their research question suggests no such prejudice. It is unlikely that this sort of study (had it been done by more experienced researchers) would have made it past the peer-review process, which acts to evaluate the experimental design and that the conclusions can be drawn from the results. As a lab project, I'd hope the instructor or research mentor would subsequently discuss the experiment and the conclusions with this student.

HYPOTHETICAL CASE: BEGINNING RESEARCHER ALLOWED TO "FAIL"

A high school student talks to their biology teacher about a science fair project. The project they are interested in is to see if regular grass can grow in sand. They plan to add grass seed to a pot of sand purchased at a hardware store. The teacher encourages them to change their research question to something along the lines of "which growing medium is better, sand or top soil?" and tells the student to have two different pots, one with top soil and the other with sand so that a proper comparison can be done. Having been a botanist before switching careers to be a high-school teacher, the teacher is fairly certain that the sand trial is likely to grow poorly, if at all, but withholds this information from the student and approves the project once the student revises the question as suggested.

Pondering Points

- Did the teacher do anything inappropriate green-lighting this study?

- Should the teacher have encouraged the student to do a thorough literature search or would that have potentially kept the student from doing it? Put yourself in the shoes of the teacher, what would you tell them to look for in a literature search?

Wrapping Up

Sometimes, particularly with beginning researchers, the process of designing and carrying out a study is more important than the "success" of the study. I am a firm believer that interest, curiosity, and inquisitiveness must be encouraged, even if the overseer/mentor already knows the answer of the study in question. Wonder cannot be taught; but it sure as hell can be cultivated. I contend that if such curiosity is fostered in younger years, as a student's knowledge base grows, they will begin to ask questions of increasing sophistication. By the time such students are into their advanced undergraduate and certainly by their graduate years, these questions will not only be sophisticated, but I'd wager that they will be novel, as well (that is, questions nobody else has asked [or at least answered] prior). Encouraging it, however, and assisting in the design of experiments in order to answer the research questions are paramount. If the teacher were to encourage measures of a literature search beforehand, a possible place to start would be to advise the student to select two types of grass: one "normal" grass and another, a type of grass known to grow on beaches, further expanding the project to different growing media and different sands.

HYPOTHETICAL CASE: STORAGE OF WASTE OF SOME KIND, FILLED WITH HARD TO UNDERSTAND JARGON

An energy company is looking for a site to dump the wastewater from their plant. As the election day approaches, they take out an ad in the local newspaper to try to assuage the fears of the community. In it, they provide the raw data from a study that investigates the safety of each of the chemicals that would be found in the waste water. The study includes a simulated ground water report that argues that, by the time the waste water seeps to the level of the water table, it will be purified to an acceptable level, showing all the calculus used to arrive at these conclusions. In addition to this high-order mathematics, the ad uses complicated terms and abbreviations without defining them.

Pondering Points

- Is it the company's responsibility to educate the public or just provide the information to them?

- Whose responsibility is it to make sure the public understands the information?

- Whose responsibility is it to make sure the public has the information?

Wrapping Up

Although unlikely to happen in such an extreme way, a great deal of confusion can result if too much scientific information is put into a publication intended for consumption by a non-science public. In fact, too much information is more likely to hurt the ability of the scientist to convey the important tenets of the discovery than to help it. Although such extreme levels of precision accompanying accuracy are surely warranted when scientists communicate with one another, settling for accuracy with far less precision is appropriate for communications with a non-science public. Generally speaking, scientists are trained to communicate with other scientists, not with a non-science public. Likewise, journalists of nearly every type are more commonly trained in the communications aspect of such a line of work, rather than in a specific field, including any particular science or science in general. Thus, to best allow for scientists and a lay public to communicate, one of two options ought to happen more commonly. In no particular order, scientists need to be better trained at communicating "like normal people." Many scientific articles are written in the passive voice and past tense. Nobody actually speaks like this in society. Scientific communications are also often rife with gobs and gobs of data. These data are meaningless to an untrained public. Journalists, on the other hand, often lack the scientific training necessary to understand the conclusions of some manner of scientific report. If the writer of the public media article intended it to be for non-science consumption, if they are themselves unable to understand the conclusions or basic tenets of the scientific report, they have no hope of rendering it an accurate way that can be consumed by the public.

HYPOTHETICAL CASE: WHEN TO GO PUBLIC

A researcher makes a major discovery regarding ancient skeletal remains identified as a new hominid species based upon visual anomalies in the remains found in a cave. They, through their university's public relations team, organize a press conference announcing the find. Upon subsequently submitting a description of their find to a high-profile journal, one of the reviewers advises DNA testing be done on the teeth of the skeleton. The editor refuses publication without this testing. When the researcher performs this test, they find the species is not new, but did have a genetic abnormality that may account for the aberrant appearance of the skeleton.

Pondering Points

- What should the researcher have done differently?

- Was the editor out of line to insist on this DNA test? Assume the DNA test confirmed a new species. Does your opinion of the editor's decision change?

- What must the researcher and their university do now?

- Would the university be justified in punishing the researcher?

Wrapping Up

Although in this case, the researcher did not necessarily commit any form of scientific or research misconduct, they were absolutely reckless from the point of view of incomplete research and also with regards to going public on such flimsy scientific support. Cases like this do not distort the scientific record nor directly harm science *per se*, because they were never part of the scientific record. No scientist worth the paper their degree is printed on would trust a popular news source over a reputable scientific source. What it does damage, however, is the public perception of science, and this gravely compromises the public's trust in science. As this trust wanes, inevitably, pseudoscience takes a stronger root, and this becomes very dangerous for the safety and well-being of the entire world. The threats caused by climate change, and the anti-vaccination movement are prime examples of the dangers associated with the public's loss of trust in science.

HYPOTHETICAL CASE: PUBLIC REPORT OF SCIENCE BY A NON-SCIENTIST

A young reporter (whose field of study in school was communications and they only took one intro-level science class) routinely browses through the journal *Science*, one of the preeminent science journals for research publications, at a local university library. In it, the reporter finds a fascinating article on a new potential cancer drug. The paper reports that, of the mice with a very aggressive strain of liver cancer that were given this drug, 7/8 survived 30 days after the administration of a high, single dose, and one died on day 8. However, kidney toxicity was observed at this high dose. Also, after sacrifice, it was found that in all of the mice the cancer spread to other organs, though they did not yet show any outward symptoms. All the mice in the control group were dead by the end of day 4. With multiple smaller doses, the survival rate was 3/8 and the surviving mice neither showed signs of kidney toxicity nor spreading of the cancer. The reporter titles their article describing this report "New Drug Cures Liver Cancer in Mice: Are Humans Next?" The reporter cites the study, but the article can only be read if you (or your institution) have an expensive subscription or pay for the individual article.

Pondering Points

- Did the reporter craft a title that was categorically false?

- Did the reporter craft a title that was accurate?

- What could the reporter have done differently?

- Do your feelings change at all if the original *Science* article is more freely available to the public?

- How do you feel about the reporter using the university library?

Wrapping Up

On the other side of the coin that is the public perception of science is how it is reported in popular media. Here, the reporter is not doing anything that could be called misconduct, to be clear. Perhaps you could argue it is misconduct to comment on something you are ignorant about, but if that's true, aren't we *all* guilty? Even that the original paper is not widely available to the public does not in and of itself constitute misconduct. Many times, if you email the author of the study, you can get a copy of the paper for free and you can easily get the author from the table of contents, which is available online even without a subscription. Note that nothing this well-intentioned reporter wrote in their title is fundamentally wrong. However, due to their inability to interpret the results properly, they were unable to provide an accurate account of the research. This too can compromise the public's trust in science if portions of this research are ever publicly reported more accurately. Thus, most ideally, reporting of scientific results to the non-science public should be done by someone trained in both communications and science. Finally, it is not uncommon for universities to make their resources available to the general public. It can easily be considered an obligation to the community that supports/hosts the university.

HYPOTHETICAL CASE: DUMB POLITICIAN

During an election cycle, a career politician with no science or engineering training rails against the laws that prohibit smoking in restaurants. They insist that it is a violation of civil liberties and that there is far less medical consensus on the dangers of cigarette smoke than the media reports. They go on to say that there is evidence being suppressed by the other major political party's activists and lobbyists that smoking actually has several health benefits, citing a conspiracy theory website known for purporting unfounded claims. They promise to reverse these oppressive laws that encroach on the civil liberties of smokers.

Pondering Points

- Is it inappropriate for a politician to utilize information found on a conspiracy theory website known to be inaccurate?

- Should public figures such as politicians or other leaders be held to a higher standard regarding what they report and/or claim regarding facts that impact all of society?

- Even if such laws are reversed, are establishments obligated to let people smoke on their private property?

Wrapping Up

Politicians and other public figures have an apparent habit of latching onto one side of a cause or another. When they do so and are uninformed regarding the topic about which they are speaking, they do so at great detriment to all of society and perhaps the world. Such discussions must be left to the experts and the experts only. People who are not informed simply must not speak publicly, especially when their words carry more weight than the average person, on a topic about which they have no training or other experience. To do so is to misuse their public pulpit. Right or wrong, many look more to such public figures than to experts for information. Perhaps it is because the public figures talk more like they (the listeners) do? Perhaps it is because the public figures may be more likely to say what they (the listeners) want to hear? Whichever the case, public figures have a way of swaying the public and they would do well to make sure they are speaking real truths before talking about matters of public importance.

HYPOTHETICAL CASE: SCIENCE-SMART POLITICIAN

A third-term mayor had a major health scare and has since been following a rigorous popular new diet. The diet has some proof of positive health benefits in peer-reviewed literature, but it is still too new to confidently know about all of its long-term effects. The mayor is open and honest about this lifestyle change and how successful it has been for them, citing the studies while also adding that more research is still being done and that nobody should consider switching to this diet without first talking to their doctor, dietician, or other health specialist.

Pondering Points

- Is it appropriate for politicians who have no training in the sciences to "take sides" of a scientific argument?

- Is it appropriate for politicians, celebrities, or other public figures to give anyone such advice or is it someone else's responsibility to decide whether or not to listen?

Wrapping Up

This is an example of someone appropriately using their public presence to impact the populace for the better. In no way is this figure claiming some sort of unfounded conspiracy nor are they claiming something that is resoundingly disagreed with by all medical professionals. They are also appropriately and faithfully sharing their experiences, stressing that there is still work that needs to be done to prove for sure that this diet is something that is truly positive. Finally, their encouragement to speak to a doctor first is exactly what any public figure should do when saying something about health to the populace. Even if this person was a doctor earlier in their career, advising people to speak to their doctor first is wholly appropriate since they may have other conditions or a health history that may make this approach inappropriate for them.

WHAT IS SCIENCE AND THE PUBLIC'S OBLIGATIONS TO EACH OTHER AND HOW DO WE FULFILL THIS OBLIGATION?

Science and the public *must* find a way to enjoy symbiosis. Most often, this symbiosis will come from the public supporting (especially financially to some extent) science and science producing things (e.g., medicines, more efficient cars, stronger materials) that the public can benefit or monetarily profit (or save) from. Sometimes, ordinarily for businesses, such support comes in the form of tax breaks, subsidized loans, and/or building or other deals. Especially in the case of academic research, the support is usually in the form of grants from federal (in the U.S., the National Science Foundation [N.S.F.], National Institutes of Health [N.I.H.], Department of Energy [D.O.E.], etc.) or state programs that are funded by tax dollars, though some private philanthropical foundations provide funding too. Decisions on what projects receive funding are almost unfailingly made by scientists, typically experts in the proposed field of study. Taxpayer money or not, this is what is appropriate and the public trusting scientists to make these decisions is part of the public's obligation to science. Only another expert in the field is truly qualified to evaluate the merit and feasibility of the project and the methods proposed to carry out the experiments described. Furthermore, a non-expert public will be more likely to make funding decisions based on reasons disconnected from science and perhaps clouded with bias based on personal or political ideology. Science has no room for either.

Science's obligation to the public goes deeper and is more complicated. This includes, but is not necessarily limited to:

- Avoiding environmental release of all research materials.

- Engaging in responsible research that will not lead to results that have a high likelihood of being threatening to Earth or civilization.

- Being completely truthful.

- Fully vetting research before communicating with a non-expert public.

Avoiding Release of Research Materials

The scientific enterprise involves a certain measure of unavoidable danger. Although enormous strides have been made, sometimes accidents happen. In the U.S., the OSHA (Occupation Safety and Health Administration) sets out the guidelines that govern safe working conditions. These conditions not only protect the worker but also by way of helping to prevent major, catastrophic accidents, help to protect the surrounding communities as well. The EPA (Environmental Protection Agency) in the U.S. and sometimes additional local authorities provides the regulations that govern how the scientific enterprise interacts with the surrounding area. For example, water waste is one particularly important point as expelling research waste can contaminate local streams, rivers, wildlife (including farm livestock), and drinking water supplies. Some areas without waste-water treatment plants may mandate that even aqueous (water-based) waste must be treated as hazardous waste and not enter the water supply. Regulations also exist that impact how much filtration or other scrubbing must be done on any exhaust fumes that are released. Biological research has very strict regulations regarding the use of pathogens in research to ensure the prevention of accidental release or contraction of a potential pandemic-causing pathogen. Other forms of research may require huge amounts of electricity or water and may impact quality of life for the immediately surrounding community. In these sorts of cases, the researchers would be good, responsible citizens to only perform these heavy resource-dependent experiments in very late hours when they would not be as impactful to the community.

To Engage in Responsible Research

Your scientists were so preoccupied with whether or not they could, that they didn't stop to think about if they should.

—IAN MALCOLM, PLAYED BY JEFF GOLDBLUM
IN JURASSIC PARK

It is for sure, a slippery slope, I believe, to tell someone that certain research is off limits. Sometimes, however, the risks of some research arguably outweighs the potential benefits. Certain of these research topics are allowed to proceed with appropriate approvals and oversight and are discussed later in more detail. For now, suffice it to say that ensuring that the research does not have predictably disastrous consequences is essential. An argument can be made in favor, however, in some cases; particularly those involving pathogens. Some contend that research into making pathogens more virulent, more contagious, or both is in fact quite necessary and truly part of science's obligation. This is justified in at least two ways:

- Natural mutations may one day lead to this more dangerous strain.

- If *we* could make it then so can *someone else*, and if they do so and then use it as a weapon, we had better be prepared to find a cure for it.

Neither scenario is implausible. And, with the increased globalization and ease of travel the likelihood that such a pathogen reaches truly never seen before level pandemic is non-negligible. Rigorous safeguards and oversight are needed and already employed that prevent the accidental or deliberate release. Understand, dear reader, that there are hundreds of known samples of smallpox in different labs on planet Earth,* kept in labs in the U.S., Russia, and the World Health Organization, despite it being otherwise eradicated.

To Be Completely Truthful

Science must always report its results with full honesty and candor. This does not mean that every single, even every failed experiment, must be reported. By no means does it mean this; only experiments relevant to (those supporting and contrary to) the conclusions need to be reported. All the evidence lending to the conclusion must be present, and must fairly represent the experiments carried out and the result(s) therefrom obtained. Only then can other experts evaluate the merit of the claims during the peer-review process. Falsifying or fabricating results does so much more than hurt science. It hurts science, for sure, by leading other researchers down potentially impossible paths as they try to utilize or otherwise further this false science. These acts also cause the erosion of the public trust in science. This has far-reaching repercussions that cause people to not use vaccines or other, life-saving medicines, or modulate their responses to environmental/climate concerns. A tipping point may well be coming in at least two of these aforementioned issues that will begin to have worldwide impact. At least one of these can be traced to one of the most infamous cases of scientific misconduct in the modern era; that of Andrew Wakefield and the MMR vaccine's "link" to autism. The study reported by Wakefield has been since retracted amid a host of ethical concerns, among them: conflict of interest and fabrication of data.† Accusations abound against climate scientists falsifying or otherwise misrepresenting data to bolster their claims of a coming

---

* https://www.cbsnews.com/news/smallpox-virus-stockpiles-wont-be-destroyed-yet/ (last accessed May 5, 2019).

† How the case against the MMR vaccine was fixed, www.bmj.com/content/342/bmj.c5347; How the vaccine crisis was always meant to make money, www.bmj.com/content/342/bmj.c5258; The Lancet's 2 days to bury bad news, www.bmj.com/content/342/bmj.c7001; Rachel Sheremeta Pepling, Chemical and Engineering News, 2008, December 15th, pg. 34. "When Controversy Shouldn't Exist" (all last accessed June 24, 2018).

calamity. Instances of industry misrepresenting pollutants, the dangers of products like Round-Up, or obfuscating the occurrence of side effects in medicines are also seemingly always in the public eye. A crime lab worker was found guilty of fabricating evidence. As a result, charges were dropped in more than 20,000 low-level criminal drug trials.* Her actions in some cases caused thousands of others to be so questionable that they no longer were able to hold up amid the scrutiny of the legal system. Clearly, the repercussions and threats of scientists lying are very real and potentially disastrous. Eventually, taken together, they push the non-science public to have more faith in pseudoscience.

All of this says nothing of the cases where science corrects itself, something that a non-science public may take as a reason to not believe anything from science. The only cure for this grave misunderstanding is an increased scientific literacy at all levels of school, something that, though very important, is beyond the scope of this book; though I do try to at least bring it up in some of the cases about how science is done.

Science Must Be Vetted by Other Experts Before its Release to the Public

Although peer review is not set up to officially check the results of a scientific study, nothing that has not gone through the peer-review process should *ever* be released to the public. Peer review too often catches genuine errors to in good faith publicly report research prior to this expert review. If errors are found by expert review after a public release, the trust of science from the public will erode once those errors are announced. When results are finally released to the public, extreme care must be taken. Scientists often communicate with one other using a style and language that is easier for a non-science public to misunderstand than to get right. As a result, some level of (sometimes significant)

---

* https://www.npr.org/sections/thetwo-way/2017/04/20/524894955/massachusetts-throws-out-more-than-21-000-convictions-in-drug-testing-scandal (last accessed May 3, 2019).

"translation" is wholly necessary. The best person to do this is a scientist who is well-trained in communication methods, though a communications specialist with extensive scientific training in the field they are reporting on could likely suffice. The public release also must properly cite the original work so that a reader can go check what the work *actually* says rather than someone else's interpretation or summary of it. If such a citation is absent, you should completely disregard the report, without exception. If you'll allow some hubris on my part for a moment, I'd also encourage you to look up the credentials of the author of the public article. If the author lacks any manner of scientific credentials, you absolutely should find the original scientific report and read and interpret it yourself. If you are reading this book, your expertise likely exceeds that of any author with no scientific training and your interpretations of the results are probably more accurate if they are not the same.

HYPOTHETICAL CASE: CORRELATION VS. CAUSATION

The adage that defense wins football games is well-known. Since it is the objective of the defensive player to tackle (interact with the other team's player in a violently physical way), a social science researcher claims that this is evidence of an increasingly violent worldwide society. They go on to cite the increase in the frequency of terrorist attacks, armed conflicts, and other violent crimes, and claim that these are the cause for the increasing focus on defense in football.

Pondering Points

- For the sake of argument, assume that violent crimes, terrorist attacks, and armed conflicts have all increased in frequency. Does it make sense to conclude that they are impacting the way a professional sport is being played?

- Is it inappropriate to find correlations between odd things if you stop short of inferring a causation?

Wrapping Up

This is a difficult to avoid temptation that many of us fail to avoid. It's also fundamentally different than being wrong. Correlation and causation are unequal. A good rule of thumb is whether or not a feasible mechanism can be at least proposed that would allow for "A" to *cause* "B." Until such a mechanism can be proposed, the two (or more) are simply correlated with each other. Even after a mechanism is proposed, causation isn't established until the experiments are done to verify the causative relationship. Naturally, some experiments that would confirm such a relationship may sometimes be too risky to actually carry out. Sometimes, a sort of surrogate endpoint is sufficient to establish a causative relationship.

HYPOTHETICAL CASE: PUBLISHING TOO SOON

A research lab observes a never-before reported chemical reaction occurring. This reaction combines two starting materials in a novel way. They quickly run a small series of similar reactions demonstrating a very impressive range of reactivity. Each reaction is run one time. After running out of the starting material, they submit a brief communication to a high-profile journal renowned for a fast publication turnaround. After preparing a new batch of the starting material, they attempt to further the results but now the reaction fails to furnish the same type of product originally observed. Attempts to repeat the results just published also fail. That the reactions worked is without question.

Pondering Points

- What should the researchers do now regarding their published results?

- If not misconduct, what may explain the observed results?

Wrapping Up

Deciding when to publish the results of a study is not always straightforward. Very rarely is there a clearly defined "end" to a research project. Rather, it is far more commonly the case that results suggest certain additional experiments be done. The field then naturally builds up in an incremental way. That being said, one can usually find "mini projects" within a larger study. Sometimes, however, in a rush to publish first, things are cut off too soon. For sure, results must be repeated before they are published. Otherwise, a case such as this one, where something inexplicable is occurring can happen. In our story, it is noted that it is without question that the reactions observed did in fact happen. However, with fresh reagents, it is not repeatable. One explanation is some sort of undetected contaminant. Note that, here, the problem is not that the researchers chose to publish part of the study; rather that they did not verify their results before publishing.

HYPOTHETICAL CASE: PUBLISHING TOO SOON V.2

A research team is investigating the effectiveness of a new chemical process. They discover that a novel reaction works very well for reagents of low reactivity, repeating their reactions in triplicate. They publish their results and insinuate that, if the same conditions were to be used with reagents of higher activity, the yields of the observed product should be at least as good, perhaps better. While they don't investigate this reaction any further, another research team attempts to independently use this reaction with a more reactive reagent in a different study and the reaction fails. A minor change to a different solvent, however, provides the originally observed chemical reactivity with slightly higher yield. Applying this change to the original reactions shows an increase in the yield of those reactions as well.

Pondering Points

- Should the original team have done more trials to verify that the more reactive cases would give more product as they had logically concluded?

- What recourse may the second team have?
- What, if any, kind of misconduct occurred?

Wrapping Up

As previously discussed, sometimes "mini projects" within a larger project are easily identified. This case here fits well into that theme. Certainly, an argument can be made that the original researchers should have pursued the additional cases, if for any other reason to "round out" their own study. However, relying on similar trends in chemical reactivity is neither bad ethics nor bad science. In fact, it is precisely good science that allowed for the new researchers to further the results of the original researchers.

HYPOTHETICAL CASE: DATA MANAGEMENT 2—LOSS OF ORIGINAL DATA

A research group publishes their results on a project developing new HIV inhibitors. Their derivatives are highly active and after more than ten years of research, they attempt to patent their work before licensing it to a start-up pharmaceutical company for further development and publishing the work in a traditional peer-reviewed journal. They find that the original data, data that is needed for acquiring patent protection from one of the earliest and most active analogs, has been lost. It is presumed that the student who generated the data inadvertently discarded the original data when they graduated. It will take at least a year to remake this derivative and repeat the testing to obtain the data again. By the time this can be done, the start-up pharmaceutical company will choose a different collaborator to purchase a license from.

Pondering Points

- Can the start-up company be blamed for a lack of patience?
- Should the group and the pharmaceutical company agree to just leave out this particular analog so as to not hold up the collaboration?
- How long should data be kept for?

Wrapping Up

There is no real established rule for a length of time that data must be kept like there is for taxes and other financial information. However, there is no reason to not keep the data indefinitely, particularly with the possibility of digitizing the results. Results stored digitally, however, should be backed up at least twice, with one of the backups being stored in another location. As the cost of digital storage continues to decrease, having additional backups should only become easier. From a data management point of view, it is also often wise to not overwrite your backups. This allows you to recover more easily from a data corruption. It is also wise for each individual to retain their own digitized back up, particularly with regard to instrumental data generated from shared instrumentation. Furthermore, on the topic of instrumentation, it must be the raw data, not only the worked-up (interpreted, digitally enhanced, etc.) data that is archived. In the case of notebook entries, typing a summary is insufficient for a digitized backup of original research data. Although there is electronic notebook software, if you have a paper notebook and plan to digitize it as your back up, it must be scanned, rather than summarized into some sort of text file. Only an original entry can suffice as a true backup of research results. Such rigor would be necessary to defend priority in court.

HYPOTHETICAL CASE: NON-COMPETITOR AGREEMENTS

A brilliant young scientist is reviewing a contract they just received from a major pharmaceutical company to be a Level 1 scientist in the neurological drug discovery division, a life-long interest of theirs. The offer contains a very generous starting salary and fringe benefits. One portion of the contract, however, gives them significant concern—the part that states that "should employment be terminated, for any reason, originating in either party, the employee is forbidden from taking a position at any company actively researching neurological drugs for seven years after the last day of employment."

Pondering Points

- Is the company within its rights to prohibit such work after termination of employment if the company is the one who severs the relationship?

- Is the company within its rights to prohibit such work after termination of employment if the employee is the one who severs the relationship?

- What should the person do?

Wrapping Up

Non-competitor agreements are not unusual, nor are they against any rules. These are particularly necessary if the employee signing such an agreement will be in possession of restricted sensitive information such as (but not limited to) trade secrets. Employers are well within their right to make sure they protect their products or customer lists (and other precious resources) from being stolen away by a competitor. Reasonable limits regarding timeline and geographic restrictions, however, must be imposed. Seven years in this sort of setting may not be unreasonable. By the time this period is over, any drug leads this worker would have been a part of will either be abandoned or through the pipeline.

HYPOTHETICAL CASE: FAST-TRACK

A pharmaceutical company is in the final stages of developing a drug to treat a disease with no other known treatments. In addition to showing a very good safety profile in humans, it has been shown in animal studies to be very effective at treating this disease. The company formally requests that the drug is granted a rolling review, which the U.S. FDA grants and allows the company to submit completed sections of important documents rather than waiting until every section is completed to initiate the review.

Pondering Points

- Is this a manner of a pharmaceutical company getting special treatment from the FDA?

- What are some of the risks associated with rolling review?

- What criteria should be met for a potential drug to be eligible for this?

Wrapping Up

This is an example of a fast-track process designed by the U.S. FDA to expedite the review of select drugs.* These drugs typically treat serious conditions or fill some unmet medical need. Here, the unmet medical need is for a nebulous "disease with no other known treatments." Although at this point, that sort of case may be rare, "serious" conditions include AIDS, Alzheimer's, heart failure, and cancer, just to name a few, according to the FDA website. The goal is to get the process of reviewing the materials to the FDA faster and to increase communication between the pharmaceutical company developing the drug and the FDA In essence, it is an acknowledgment by the FDA that some potential drugs may *deserve* some special treatment so that people can benefit from them being available more quickly. It is important to note that this process does not allow for a shortcut in the way of less data gathering regarding safety and efficacy by the sponsoring company. All the typical data must be collected. What actually happens is that the increased communication between the company and the FDA allows for insufficiencies to be found earlier so that they do not hold up the process late in the game.

HYPOTHETICAL CASE: BREAKTHROUGH THERAPY

A pharmaceutical company has developed a drug that is the first new one in decades to treat a disease. During trials, it is first

---

* https://www.fda.gov/patients/learn-about-drug-and-device-approvals/fast-track-br eakthrough-therapy-accelerated-approval-priority-review (last accessed May 5, 2019).

demonstrated as safe (in phase I) and then found to be far more effective than all the older drugs used to treat the same disease during phase II trials. The FDA approaches the pharmaceutical company, contacts the company, and encourages them to pursue a breakthrough therapy designation; something that will afford the company all of the fast-track designation features along with other intensive drug-development guidance.

Pondering Points

- In whose interests is the FDA acting in to make the recommendation to the pharmaceutical company?

- What threshold(s) should be reached by a drug candidate to earn such benefits?

Wrapping Up

Breakthrough therapies, drugs that may demonstrate substantial improvement over available therapies, are afforded the same benefits as drugs that are granted fast-track designation.* The goal in both is to bring a new drug to the market faster by expediting (but not cutting short the experiments) the process to evaluate a drug. A drug that either shows significant increases in efficacy (i.e., survival rate or another surrogate endpoint) or significant reductions in either serious or non-serious side effects would qualify.

HYPOTHETICAL CASE: PATENTING F GENES

A genetic research team synthesizes a gene sequence that is able to reverse genetically caused blindness. The gene sequence they synthesize is identical to a healthy gene sequence in sighted individuals. They attempt to patent their gene sequence and the patent is rejected. A separate patent that describes the novel method incorporating the genes into a patient is accepted.

---

* https://www.fda.gov/patients/learn-about-drug-and-device-approvals/fast-track-breakthrough-therapy-accelerated-approval-priority-review (last accessed May 5, 2019).

Pondering Points

- On what grounds could the patent describing the genes be rejected?

- On what grounds could the patent describing the method incorporating the genes be accepted?

Wrapping Up

The patent that describes the genes is describing something that is found in nature. Generally (as of early 2019 and since a 2013 Supreme Court decision*), a product of nature such as a gene sequence is not patentable in the eyes of the U.S. Patent and Trademark office. Even the use of genes is generally speaking not patentable, but a method that does something with the genes can be, particularly if it is different from the other gene-editing techniques (such as CRISPR) currently in use. Novelty is important in patents. Synthesizing a gene sequence, no matter how difficult the synthesis, is insufficient to achieve the novelty necessary to be awarded a patent.

HYPOTHETICAL CASE: THE LABORATORY NOTEBOOK

A research team has a policy of spending the final hour of each work day using a buddy system to review and sign each other's laboratory notebooks, each party of a pair keeping a copy of the doubly signed pages. All materials or data collected are visually verified at the end of each day; and at the end of each week, the laboratory archivist creates a digital copy of all the signed notebooks pages, returning the paper copies to the researchers each Monday. The digital copy is stored in two locations, one of which is off-site in the cloud behind a password-protected firewall.

---

* https://www.sciencemag.org/news/2013/06/us-supreme-court-strikes-down-human -gene-patents (last accessed May 5, 2019).

Pondering Points

- Is this level of redundancy overkill? At what point, if any from the point of view of financial (e.g., patent-protection) stakes, is this level of redundancy appropriate?

- What complications may come from such rigid protocols?

- What benefits may come from such rigid protocols?

Wrapping Up

Data management is critically important. This is particularly true any time the work being done may be patented or otherwise marketed. Being able to prove inventorship may be necessary in such cases and these sorts of documents would be necessary to win any court decision. Although it may seem excessive, having the notebooks witnessed is important as it adds to the credibility of the record as, under this circumstance, it is not just the author but another expert who is verifying that the work was done and that the results were recorded at the point of the work being done. While some may claim that this is something that gets in the way of getting work done, I would argue that such a stance is nonsense, as this is part of the work. Regarding the rigorous backup protocol, if someone can afford the storage, there is no reason to not have an extensive backup system. Certainly, it may get to a point where it is difficult to manage. However, if it is done right, it will be quite easy to keep everything in order and accessible by the interested parties.

HYPOTHETICAL CASE: DATA BACKING UP

A research lab specializes in the use of a specialized instrument for spectroscopic analysis. The lab serves as a hub for the entire university. All the data acquired is stored on a single computer in the lab. After five years of service, a major storm causes an electric surge that destroys the computer on which the data is stored. As a result, years of original data are lost.

Pondering Points

- What complications exist now for any of the previously published research using data gathered from this instrument?

- What complications exist now for any of the as yet unpublished research using data gathered from this instrument?

Wrapping Up

This sort of scenario is precisely why data must be backed up. With all of the original data destroyed, if anyone's results are ever called into question, their paper may have to be retracted since they would not be able to provide the original data to defend their results. Better still than just backing up the data is to keep at least one backup off-site. Whether that off-site location is in the cloud, some sort of on-campus server, or a memory stick, at least one copy of the backup should be in a different building. Imagine if, rather than an electrical surge, there was a fire, flood, tornado, earthquake, or some other natural disaster that damaged large portions of the building. If the back-up is also housed in the building, it too may be damaged and therefore, worthless.

HYPOTHETICAL CASE: SLOPPY DATA COLLECTION INTO NOTEBOOK

A brilliant but eccentric graduate student often records their experimental records on small pieces of paper and at the end of the day transfers them into their notebook. They argue that sometimes the science is just moving too fast and that they are able to get more done if they do a lot of work, take brief notes, and enter everything into the notebook at the end of the day based on those brief notes.

Pondering Points

- Under which conditions is it more likely that a detail will be forgotten or mistaken when being entered into the notebook, those described in this story or conditions where the student records everything directly into the notebook as it is done?

- What consequences may occur if there are errors in the laboratory notebook?

Wrapping Up

Best practices mandate that research results are recorded directly into the notebook—the official document that details an experiment—as the data is generated. Everything you do and every observation you make is technically data, including the quantity of any reagents you used and the length of time any process was allowed to run. If you do not record it at the time of the generation of the data, there is a great risk of making a mistake in recording the data. If there are errors in the recording of the data, or if one of the little notes were lost, then what was actually done could never be accurately reported, eliminating any chance of it being repeated. Notes may also come out of order or be lost, further enhancing the possibility that errors in transcription may occur.

HYPOTHETICAL CASE: WHAT TO DO WHEN MODEL SYSTEMS FAIL

A synthetic organic chemistry lab is attempting to develop a new synthetic process. In their study, they start with the analog that should be the most reactive and successful example. Several attempts and conditions fail to give a very high yield of the expected products. In a gamble, they decide to try the less reactive examples their actual system comprise. To their surprise, these examples work much better than the model system did. In further investigating the reaction, they discover a new effect that makes the more reactive starting material work poorly during this reaction.

Pondering Points

- Under what conditions may it be still worth it to do model studies?

- Is it bad science to not perform model studies with simple or more reactive systems first?

- Is it unethical to not perform model studies with simple or more reactive systems first?

Wrapping Up

Model studies are commonly done when researchers are unsure whether a reaction will work. This is particularly true in instances where the reaction is in the late stages of a multi-step pathway and utilizes a precious reagent that days or more were invested in preparing. Rather than perform this reaction on the precious reagent, it is wiser to test the conditions on simpler, less precious systems first. In principle, these systems should be as similar as possible to the real system to minimize the chances of being fooled by the success of the model reaction. Understand well, however, that sometimes the model fails to predict success and also sometimes fails to predict failure. If you're wondering "why bother," you're likely not alone. They are still the most cautious action to take and likely will continue to be a standard approach for the near future.

HYPOTHETICAL CASE: NEW RESEARCH PROVES OLD RESEARCH WRONG

For years, it has been assumed that a particular plant produced an important anticancer compound as a secondary metabolite. Recently, new research has found that the actual source of this compound is a mold that grows symbiotically on the roots of this plant. The plant then uptakes the compound, which then concentrates in the leaves; the compound also has properties that ward off attacks from a mite that otherwise ruins the plant. The previous studies only investigated the leaves of the plant and the new study identified the mold's presence by first analyzing the soil around the plant, and later identifying it as coming from the roots of the plant into the surrounding soil.

Pondering Points

- Did the initial researchers perform either bad science or bad ethics in not analyzing the whole plant?

- If not for finding the mold in the soil, it may never have been found on the roots. Does this impact whether or not the initial researchers did anything wrong?

- Does the source of the compound matter?

Wrapping Up

Progress often involves or at least is inspired by the generation of new knowledge. Sometimes, misconceptions are resolved by this generation of new knowledge. This can particularly be the case with regard to instrumentation. Instruments become increasingly sensitive, accurate, and precise routinely. As our ability to analyze things increases, so does our ability to understand them. If current instrumentation is only sensitive enough to show one entity, it is easy and logical to infer that any phenomena observed are due to that entity. However, as analytical techniques increase, we may eventually observe that there is in fact more than one entity present. It now becomes necessary to explore *which* entity is responsible for the phenomena and it may turn out to be the entity present in lower amounts. In such cases, nobody has done anything wrong. The best and most logical conclusion was based on the data that was available at the time.

HYPOTHETICAL CASE: DIVIDING OF WORK BETWEEN COWORKERS

A principal investigator (PI) oversees a very large research group. There are a few diverse ongoing projects and the students are sorted into projects based on their interests. Each of the projects has a postdoctoral research associate that serves as the team leader. There is a rigorous schedule of mini-meetings and full group meetings that the PI uses to keep themselves aware of the work being done by the entire group. Each of the group members is rotated to different parts of the same project over their tenure in the lab to expose them to as many techniques and parts of the project as possible, their proceeding through the phases of the projects determined by the results they obtain. Any grants that are written for each project are fully reviewed by all of the researchers on the project and all are also expected to contribute to the writing of the grant.

Pondering Points

- Is the PI shirking any responsibilities by allowing a post-doctoral research associate the opportunity to serve as a project leader or are they as responsible for the development of this researcher as they are for the development of the student researchers?

- Would it be better to rotate the students through all of the projects to diversify their education as much as possible?

- Is the amount of meetings overkill and inhibitive of the work being completed?

Wrapping Up

In a research setting, the dynamics of a research group can sometimes be difficult to navigate. The dynamic outlined here is not unusual, but that doesn't mean it's typical. Even the tactic of installing a post-doctoral researcher as a project leader is not inappropriate. Many post-doctoral research associates go on to eventually lead their own labs. On some level the working relationship between the PI and the post-docs (as they are called) is one that is not just expected to yield research productivity but also additional training. This training (in the best of scenarios) also includes grant writing; extending that experience to the graduate students only further enhances *their* education.

HYPOTHETICAL CASE: KEEPING IDEAS FOR YOURSELF

A postdoctoral research associate is beginning to prepare their application for a tenure-track position, the next logical step in their career and goals. While doing some preliminary work to prepare a paper for their current research, a new, only tangentially related direction becomes clear to them. If they are correct, it would be easy to explore and likely complete this direction prior to leaving for their next position. Rather than explore it now, they write up the idea as their proposed future research in their application.

Pondering Points

- Does the worker owe it to their current employer to follow this idea now, rather than wait for their own career?

- Assuming the idea works as well as anticipated, will any manner of co-authorships be due to the current research mentor/employer?

Wrapping Up

Generally speaking, you do not owe your employer your best ideas that are unrelated to the job you are currently doing. Certainly, if an idea is directly related to your current work, you probably should share it, but even then, you are not obligated to. This is a far riskier endeavor, however, than the former scenario. This is because the more closely related to your current work the idea is, the more certain it is that you will have well-established competition in completing the work once you are working independently. Therefore, it would arguably do more good for your career to pursue such ideas more immediately, rather than hold them off for yourself later. Understand well, however, that it would be crossing a line if you were to start work on your idea while still employed someplace else, unless pursuing such opportunities are expressly permitted.

HYPOTHETICAL CASE: TRADE SECRETS

A company discovers a mixture of chemicals that fresh-picked fruits can be sprayed with that at once inhibits mold growth, prevents attack from insects, and keeps the crop from "getting funky" in the refrigerator. The mixture is also demonstrated to wash off of the fruit to levels below the detection limits of the most sensitive analytical instrumentation available today by simply dunking the fruit in three successive bowls of fresh water. In the patent application and all public communications, they refer to the compounds using a coded name and never reveal their chemical structures or the relative percentages of each compound in the mixture. Rigorous and verified safety data are presented that detail the physiological properties of these compounds, both individually and in the relevant relative percentages in animal studies.

Pondering Points

- Should the company be forced to reveal the structures?

- Should the company be forced to reveal the relative percentages of the components?

Wrapping Up

Companies are permitted to withhold certain aspects of their product from detailed descriptions in order to prevent them from being duplicated. One of the most famous examples is the Coca-Cola recipe. Coke has worked very hard for a long time to protect its recipe. It is not even a foul if workers in a company are required to sign non-disclosure agreements that prohibit them from sharing such protected information. Sometimes, you may even be prohibited from taking a subsequent job at a competitor during some defined period of time. In short, you are allowed to protect your investment and product, even to the extent of leaving critical data out of a patent to prevent it from being forged.

CHAPTER 2

Human Subjects

WHAT ARE THE RULES FOR HUMAN SUBJECTS RESEARCH?

Most, if not all, of the developed world has some significant measure of protection for human research subjects. These protections are largely based on international laws that stem from times of war and some countries or even localities employ stricter regulations yet. Nevertheless, violations of these kinds—that is, violations of protocols established to protect human subjects and, in a later chapter, animals—are wholly different from fabrication, falsification, and plagiarism; the traditional actions considered around scientific misconduct. Violations against human and animal research subjects can result in loss or maiming of life rather than harm an institution such as science. Thus, enormous care must be undertaken to minimize the likelihood of such an outcome with regard to all manner of live research subjects.

Rules For HRS

Although to some extent, many countries have agreed on a base set of rules, small differences exist. Herein, only the general guidelines will be discussed. It is *essential* that you investigate your local guidelines and your institutional review board (or its equivalent committee for human research subjects' oversight) before undertaking any work. Since such a body would be empowered to prevent your research from even starting, I urge you to involve them from the planning stage on, so that you do not waste time planning something that will not be approved. There also may be training you must complete before you can start work; another reason to make sure you involve the committee as soon as possible.

One of the unwavering requirements is informed consent.* All research (with one notable exception) requires some manner of informed consent where the research subjects are made to understand the research goals and risks. Education research is usually an exception, but only the Institutional Review Board (IRB) or its equivalent or designee, can rule a research project (even an education-based one) exempt from informed consent. Also, the research must have benevolent or altruistic goals. As an extreme example, you would not be able to, nor would you want to (I hope) research how much gasoline someone must drink in order for it to be fatal.

Importantly, the research needn't be risk-free, the risks simply must be appropriately outweighed by the predicted or intended benefits. Research subjects are also allowed to withdraw from the study at any time and for any reason. Related to this, the researchers are forbidden from coercing or otherwise forcing a participant to enter or remain in the study, though participation *can* be compensated. For this reason, some groups are said to be vulnerable and require extra levels of protection during the informed consent process: Children, the extremely poor, illiterate persons, the mentally disabled, and prisoners all fall into this category. This is because they are either deemed unable to understand the risk, or are more susceptible to (in some cases self-) coercive measures. It is also essential, even in education research to preserve the anonymity of all participants, at all times; even after the study

---

* https://www.hhs.gov/ohrp/regulations-and-policy/guidance/checklists/index.html (last accessed May 5, 2019).

is concluded. No identifying information that would permit one to track a result back to an individual can ever be shared. This is even true for education research where federal laws protect the students' privacy, just like medical records are protected by law.

Although it may seem like it should be obvious, the individuals doing the research must be qualified; this includes having the expertise to identify when the study may become a threat to its participants. You would not expect a chemist to oversee a psychology experiment that observes couples arguing, nor would you sit idly by while the janitor started your IV in the emergency room. The IRB also ought to maintain oversight throughout the project to ensure that the safeguards are working and that the protocol is being followed.

A current gray area, and one that differs a great deal from country to country, is genetic research. There are calls for a moratorium on any research that alters the germline of humans.* In the U.S., in fact, such research is illegal though other genetic research is more variable from country to country. Recently, the World Health Organization (WHO) has begun to consider modifying its stance on the genetic modifications of embryos in response to recent genetic research in China.† He Jiankui recently reported the birth of babies whose genetic makeup was edited using CRISPR technology. The goal of the editing was to make them more resistant to HIV. The details are not important to this book. By the time this book is in print, the WHO may be done with their revision... Even within some countries, different localities may have slightly different rules. Furthermore, there may also be a difference in what is permitted by law and what public funds can be devoted to. Non-publicly funded work could then only proceed with private (and typically lesser) funding, provided it is legal. The solution for knowing the rules in your location is for

---

* https://www.washingtonpost.com/science/2019/03/13/nih-top-scientists-call-morato rium-gene-edited-babies/?utm_term=.b6adbb7b1a48 (last checked May 5, 2019).

† https://www.smithsonianmag.com/science-nature/whats-new-whats-not-reported-birth-crispr-babies-180970935/ (last accessed May 5, 2019).

the interested reader to research their own locality's rules before planning a project. International collaborations add an additional layer of complexity and, generally speaking, the more stringent regulations must be followed, even if the research actually being done happens to be in the less strict country. The researchers in the stricter country are not allowed to operate under anything but the regulations that their country mandates.

HYPOTHETICAL CASE: PRISONERS IN TESTING

A pharmaceutical company is recruiting healthy volunteers to test a new drug on during phase I trials. They approach a nearby minimum-security prison to offer prisoners a stipend and increased recreation time each week for participating in this study. All the necessary oversights from the IRBs are regularly performed and the study is approved. Meanwhile, they also recruit other volunteers that are not in jail. These free volunteers are also offered a stipend, but it is twice as high as the stipend offered to the volunteers in jail. These out-of-jail participants are not offered any manner of recreation time.

Pondering Points

- Is the disparity in pay fair to the prisoners, or is the inclusion of recreation time enough of an added compensation?

- Is the inclusion of recreation time into the "payment" to the prisoners potentially something that would compel a participant to not withdraw from the study if it causes suffering of some significant kind?

Wrapping Up

Prisoners represent one of the protected classes of research subjects as it relates to human subjects in research. Whether or not recreation time is worth the disparity in pay may well be in the eye of the beholder. As onlookers, we must trust that the IRB and those both overseeing and carrying out the study will be forthcoming and honest about the dangers associated with any study involving humans. Anything that gives prisoners some relief from the stresses of prison will potentially impact their willingness to stay in a study that may otherwise threaten them. That being said, to deny them the benefits of participating in such studies can be construed as a violation of their constitutional rights. Thus, some manners of protection have been developed to allow for their ethical participation in such studies.

HYPOTHETICAL CASE: HUMAN SUBJECTS TESTING A DRUG

A pharmaceutical company is evaluating a new potential anti-HIV drug *via* approved protocols. During animal testing, two side effects were observed in a small number of animals that were deemed serious: Sleeplessness and diarrhea. These side effects were conveyed to all of the healthy volunteers in the phase I trials. During the phase II trials involving early diseased individuals, the side effects are once again reported, despite their not being observed during the phase I trials. After seeing success in the forms of both safety and efficacy during the phase II trials, the pharmaceutical company begins phase III trials, increasing the pool of patients to individuals with advanced-stage HIV. Some mild side effects are observed during phase III trials and all results, including side effects and efficacy, are reported to the FDA periodically throughout the evaluation. All the data was ultimately collected, collated, and submitted to the FDA by the company carrying out the study. After nearly a decade of work from start to finish, the FDA approves the drug for use in treating early- and advanced-stage HIV.

Pondering Points

- Is it appropriate for the pharmaceutical company to be so involved in the collection and evaluation of data for its own drug?

- Is it appropriate to give an experimental drug to healthy volunteers?
- What side effects, if any, are tolerable?
- Does the severity or morbidity of a disease impact what side effects should be tolerated?
- Could the same side effect be tolerable in one treatment setting but intolerable in another?

Wrapping Up

The pharmaceutical company sponsoring a drug through the approval process is always the entity that actually carries out *and pays for* the study. They are required to report all of the results to the FDA (in the U.S.) and significant independent oversight is also provided. Doctors, not the employees of the company, administer or dispense the medicines to patients. Pushing such costs onto tax payers would cripple the system. Also, the discovery and optimization of a drug is very much a trial-and-error, iterative process. Only by carefully designing medicines for safety, efficacy, and bioavailability (how long it lasts in the body, effectively) can a successful drug be found. Furthermore, a variety of factors influence which targets are ultimately brought forward for full testing. One of the seemingly inappropriate, but entirely rational factors is the cost of making the candidate. If a particular candidate cannot be synthetically prepared in high yield using current chemical technology, the cost to prepare large quantities needed to test (much less deploy it as a chemical treatment) becomes so great that it is impossible to recoup these costs by consumer purchases. This is not some example of capitalistic greed harming the consumer. The raw materials, the synthetic chemists, and all the infrastructure (especially safety) resources must all be paid for. If current technology can only create a process that enables a drug to be made such that the cost to the company making it is thousands of dollars per gram, it will never be a drug, no matter how great the activity.

HYPOTHETICAL CASE: PSYCHOLOGICAL STUDY USING HUMAN SUBJECTS

A research team is investigating if losing streaks, winning streaks, and distractions influence someone's willingness to take risks during online gambling. As part of their study (which uses fake money), they withhold certain pieces of information (of the conditions) regarding the study from the informed consent. Their plan is approved by their I.R.B. The computer program they are using is set to three different conditions: 1—"fair play"; 2—stacked in the player's favor for winning streaks to be more likely; 3—stacked in the house's favor for losing streaks to be more likely. The participants are unaware of these three settings. The justification for withholding it is that knowledge of the program being so would interfere with the study. As detailed in the informed consent, distractions occur—although the participants were not told what the distractions would be. The distractions include: a couple masquerading as participants who begin fighting; a fire alarm going off; and hostesses offering drinks of the participants' choice (including alcoholic beverages). After participating, during the debriefing, participants are informed of the different playing conditions; they are also informed that the fighting couple are actors. Anonymity is ensured for all participants and they are financially compensated for their time.

Pondering Points

- Is it *ever* ethical to withhold information as done in this study or should the participants be trusted to not be so influenced?

- Why is anonymity important?

Wrapping Up

In some instances, and with the appropriate approvals and oversight, it is appropriate and even necessary to withhold some pieces of information from research subjects. In all cases, this withholding must pose minimal to no risk and this can only be evaluated by the relevant review committees. In a case like the one described here, the behavior of the participants would almost certainly be influenced by knowing that the computer may be programmed to yield winning or losing streaks. Thus, in order to truly measure if such streaks impact their behavior, the information that the streaks are programmed to occur must be withheld. After the study is over, a debriefing done by the researchers must clearly explain this experimental design and all of the other pieces that were withheld from the subjects. This study should also arrange and pay for the travel of all participants who consumed alcohol.

HYPOTHETICAL CASE: MANIPULATION OF HUMAN SUBJECTS TO STAY

An IRB-approved study is being done to investigate the emotional response to one of twenty sensitive themes in a movie, including child abuse. The sensitive themes are randomly chosen in each case. All individuals are to be recorded during the movie. One of the subjects is immediately uncomfortable as their movie starts; it is one of the movies with child abuse; this subject was abused as a child and has never told anyone. They indicate shortly after the movie begins that they are unwilling to continue and wish to withdraw from the study. The researchers indicate that in order to withdraw, an individual must indicate why they are withdrawing so that at least some data can be gleaned from their participation. This is a departure from the approved protocol. This interaction is caught on video. Rather than share their secret, the individual changes course and stays in the study. Later that night, they commit suicide. In a letter, they out their abuser and cite being forced to stay in the study as the trigger for their suicide, lamenting that they were not being allowed to withdraw. Their family then sues the researchers.

Pondering Points

- Should the family be awarded damages?
- Could even withdrawal of a subject in a case such as this be considered as a valid data point?

- What should the researchers have done differently?
- Should the subject have done something differently?

Wrapping Up

For sure, this is a deliberately heart-wrenching and extreme example. The final question treads closer than I am comfortable to victim blaming; however, part of discussing this topic necessitates uncomfortable topics to some extent. For certain, in this case it is hard to find fault in the victim's actions. They were twice a victim—first by the person who abused them all those years ago and now by the researcher—though the abuse this time is different. However, it is precisely to safeguard against such damages that persons are permitted to withdraw from any study and for any reason. It is not permitted to force such a participant to provide justification for their decision to withdraw. To try to manipulate someone to stay in a study is in and of itself abuse and is not ethical. Whether or not withdrawal could be considered a data point may be up to how the datapoint is then used. It is unlikely to be able to add to any conclusion, but evaluating if any one topic had more or less withdrawals would potentially be valid.

HYPOTHETICAL CASE: HUMAN SUBJECTS UNNEEDED SUFFERING

A drug is entering clinical trials. The drug is for late-stage pancreatic cancer. In animal studies, it was found to be 56% effective at curing the disease. There were significant side effects though, including extreme sensitivity to touch and irritability. In healthy humans, additional side effects were sensitivity to light to the point of not being able to go outside, hallucinations to the point of not being allowed to drive, and an insatiable sexual appetite. For some perspective, the one-year survival rate of pancreatic cancer is 20% and the five-year rate is 7%.*

---

* http://pancreatic.org/pancreatic-cancer/about-the-pancreas/prognosis/ (last accessed May 5, 2019).

Pondering Points

When the option is death, are such side effects tolerable when in other instances they would not be?

Wrapping Up

The mortality of a disease or the pain and suffering associated with it greatly impact the side effects that are tolerated for treatments. For example, hair loss is tolerated as a side effect for cancer treatments. If this were a side effect for a treatment for a headache, it would be completely intolerable. Ultimately, the FDA decides if the side effects are too great to warrant approval. The pharmaceutical company can also opt to withdraw the drug if side effects are observed after the drug is deployed for use. A company may do this if they feel threatened by the bad press that some side effects are receiving (from a public relations viewpoint), or if they are concerned that the FDA may withdraw approval.

HYPOTHETICAL CASE: HUMAN SUFFERING GOOD JOB

A drug candidate showed unprecedented efficacy in resolving infection with a disease *in vitro*. In animal studies, there were no observed side effects and the efficacy was on par with the *in vitro* observed results. However, during Phase I clinical trials in healthy humans, a large percentage of the participants suffer from a range of side effects, including hair loss, osteoporosis, and permanent blindness and numbness in the hands and feet. The drug is then subsequently abandoned.

Pondering Points

- The disease is deliberately not mentioned. Are there any diseases that are so bad that curing them would be worth any of these side effects?

- Did the drug sponsor (or the FDA) overreact by withdrawing the drug so quickly from consideration?

Wrapping Up

In this case, the list of side effects is bad enough that almost no disease's cure would be worth these side effects. Side effects this severe would warrant ending a study even if they were only observed in a small number of patients. If common traits among the participants who suffered these effects could be found such that a group or groups of patients that are safe to administer the drug to could be identified, perhaps it would earn re-entry into trials, but the activity profile would have to be truly ground-breaking and exceptional to warrant taking the risk to prove the safe group right.

CHAPTER 3

Research With Animals

WHAT ARE THE RULES FOR RESEARCH WITH ANIMALS?

Setting aside any kind of argument regarding which life is more important (humans vs. anything else), research that involves animal research subjects also requires a great deal of oversight. Once again, loss of life is possible as is the terrible suffering of the research subjects. A noteworthy complication with animal testing is that they are (of course) less capable of communicating than a human. They (animals) cannot tell you they have a horrible headache or stomachache. They cannot tell you that they are seeing or hearing things and they cannot tell you about this weird dream about a talking fire hydrant over a spot of tea. This complicates performing animal testing in a way that ensures the respect and safety of this life. And this is at least part of why a trained veterinary medicine expert is a mandatory member of the team, as they are better trained to identify stress in an animal.

Animal testing is at least as regulated as human testing, particularly in the U.S. Notably, invertebrate animals (including all insects) are not covered by the current regulations. Unlike human subjects that verify exemption from informed consent, no such verification is required for these research subjects. An exhaustive coverage of the rules is inappropriate—it could fill a book in and of itself. Understand, however, a few general guidelines. First, like with human testing, animal testing must be done by qualified individuals. A veterinarian must also be involved with the study. Second, although certain animals are required to have certain amounts of exercise and/or social time, some forms of research—such as those investigating contagious diseases—would receive an exemption from at least the socialization. Third, there are certain standards of care/cleanliness that must be provided. Animals cannot be in filthy, waste-laden areas. They also must be kept in enclosures that are appropriately sized. Fourth, appropriate, minimal-suffering euthanasia methods must be employed when killing an animal. Fifth, animals shall not be forced to undergo persistent suffering. This summary, however brief, should provide at least a baseline guide. For more information about local guidelines, consult with the appropriate committee at your institution.

All of this, however, may be coming to an end. Serious efforts and calls have been made to end animal testing. Methods are being developed that purport to make animal testing unnecessary. Until very recently, no suitable alternative has existed to replace animals as a preliminary screen to evaluate a drug's safety and an indication of its efficacy against at least a model of human disease in a living system that is more dynamic and complicated than a well of cells.* For sure, some—and not just activists like the People for the Ethical Treatment of Animals (PETA)—will welcome this change with great enthusiasm. Equally certain, however, is that there will be others who will warn that the loss of this informative step will cause human

---

* https://www.niehs.nih.gov/health/topics/science/sya-iccvam/index.cfm (last accessed May 5, 2019).

suffering in future clinical trials. Time will tell who is correct on this matter. Unfortunately, by the time we find out it is necessary (assuming it is), it will be too late to undo the harm caused.

HYPOTHETICAL CASE: INSECTS DON'T COUNT

A team of researchers is running a project where they grow several large colonies of earthworms under a variety of different soil, food source, insecticide, fertilizer, and temperature conditions. After the set time periods (several months), they count the number of worms in each set, weigh them, and then blend them into an aqueous heterogeneous mixture using a kitchen blender. After filtering, they run an isocratic extraction to isolate secondary metabolites on the aqueous mixture and any solid matter. Several leads with interesting biological activity are identified and the work is published in the journal Nature.

Pondering Points

- There is no mention of an IRB-approved protocol. Assuming this is not an oversight of the author of this book (it is not), what do you think about this?

Wrapping Up

As gruesome and barbaric as this may sound, a study such as this is exempt from any manner of oversight. Current regulations hold that insects and all invertebrates do not rise to the level of needing oversight. You can literally do anything your conscience permits you to do to an insect. Don't misunderstand, some behavior may cross the border to sociopathy. But from the point of view of *legitimate research*, insects do not require oversight. Without a valid scientific goal, however, something like pulling the legs one by one off a cockroach is wrong and I would be surprised, almost impressed, in fact, if a scientific reason for *that* can be conjured.

HYPOTHETICAL CASE: ANIMALS LOST

A research institution is running at its maximum allowed capacity regarding the number of animals in the aging facility. All protocols for the experimental use of animals and the containment of deadly pathogens are approved and strictly followed. Several of the animals are genetically engineered for a specific disease trait the institution is researching that makes them more susceptible to the human form of Ebola, including a batch with the disease. One night, during a particularly bad thunderstorm, lightning hits the generator resulting in a terrible fire that kills many of the animals and results in the release of most of the rest of the animals into the wild, including those made more susceptible to Ebola. An Ebola-carrying batch is unaccounted for and assumed but not confirmed to have been released too. All of the employees escape, saving whatever animals they could fit in cages and get out with.

Pondering Points

- What, if anything, could have been done to avoid this loss?

- What should have been done about the Ebola-carrying animals that appear to have escaped?

- Does the institution have an obligation to the public's safety now?

Wrapping Up

Natural disasters happen, however unfortunate they are. Inevitably, research facilities are caught in the crosshairs of such calamities. The loss of life at animal research facilities is only the tip of the disaster iceberg. As seen in this case, even in spite of rigorous controls being adhered to, the accidental release of dangerous pathogens (in this case) or chemicals can occur. It is unreasonable to expect anything to be disaster-proof, though it doesn't mean we can't do our best. Somewhat recently, some of these issues came to the forefront with major hurricanes or other storms in coastal cities.* Though in those cases, it wasn't reported that any potentially dangerous animals were released.

HYPOTHETICAL CASE: CHIMERIC ANIMALS

A research team is attempting to investigate how an animal adapts to changes in its body. Their purpose is to investigate whether animal body parts could be transplanted similar to the way organs are done. It is their intent that, if successful in animals, it may be possible to replace limbs in people using other animals rather than prosthetic limbs. In their study, they replace a pig's foot with that of a live gorilla. After allowing the wound to heal, the pig is observed to have good mobility and it is also found that it reacts to stimuli in the foot such as pinpricks, tickling, heat, and cold.

Pondering Points

- What concerns do you have about this research?

Wrapping Up

Although there are other issues with this study, particularly what happened to the gorilla, the use of chimeric animals, that is, animals that are one part one animal and another part different animal is done. In fact, in 2016, the NIH revised restrictions on human–animal chimeric research from the point of view of allowing human stem cells to be added to animal embryos. This is still a matter of significant discussion.† Studying human development, modeling diseases, or limb/organ regeneration are some of the goals for this research. This field of research is likely to intensify and continue to change in the coming years and decades. Currently, the review and oversight on this mode of research are tight. However, if its safety is established, it is likely to become looser, but that is just a prediction by me.

---

* Yahoo! MAKTOOB News article, "NY University faces growing criticism after Sandy kills lab mice" (last accessed August 5, 2013) and NPR article, "A tale of mice and medical research wiped out by a superstorm" (last accessed March 6, 2016).
† Levine and Grabel, *Stem Cell Research*, **2017**, *24*, 128–134.

HYPOTHETICAL CASE: CHIMERIC ANIMALS—GENETICS VERSION

A genetics research lab believes it has identified the gene sequences that are responsible for the creation of a human voice box and also the gene sequences responsible for developing the Broca area (the area of the brain responsible for speech). They similarly believe they have found the area of a dog's genome that is responsible for developing its equivalent of a voice box. Using genetic engineering technology, they replace the gene sequence of the dog that generates its equivalent of the voice box with that of a human and subsequently add to the dog's genome the sequence responsible for developing the Broca area.

Pondering Points

- What could possibly go wrong?

- Why should or shouldn't this research be done?

- What disease or human condition could this possibly help to solve?

Wrapping Up

It is unlikely that this research would pass the current muster of the review from the NIH to be approved. This sort of study would not address any diseases or other human conditions. Although it may say something about language acquisition, there is nothing here that could not be learned by other means. It also brings about a wide range of other ethical questions; specifically, how we should respond if the dog acquires an enhanced ability to communicate. While on the one hand, better understanding of non-human animals would be beneficial, what level of enhanced rights would be due to such a dog? If this dog is then able to vocalize that it is feeling lonely and bored, is it suddenly animal abuse to not play with the dog until it indicates it is no longer bored? If the dog says "this food is disgusting, why do you keep buying it, I don't want it anymore," are you abusing your dog if you don't buy different food? For sure, these are somewhat humorous hypotheticals, but the ethical questions behind them are genuine. If an animal, any animal, is better able to voice displeasure or discomfort, is there an obligation to (at least try to) resolve the discomfort?

HYPOTHETICAL CASE: ANIMALS' UNNECESSARY SUFFERING

As part of a class, a university student is doing research on how mice respond to different chemicals. They place varying amounts of a random chemical they find in the chemical stockroom in the feed or drink of the mice, varying the concentration of the chemical. They caught the mice around campus using live traps. At all concentrations, the mice die within a few days of consuming this tainted food and drink. The mice exposed to the highest concentrations clearly suffer as they vomit blood and convulse during their final minutes of life. They reason in their report that "Since everyone thinks mice are icky, it's not really a big deal that they all died."

Pondering Points

- What should the student have done differently?

Wrapping Up

Mice, as gross and as typically associated with disease and filth as they are, are vertebrates. This means that oversight from some sort of IRB and animal subjects committee would have needed to be involved and virtually no committee would approve the use of animals in this way. This is fundamentally different from testing the safety of a drug on animals in that there is absolutely zero benevolent goal behind testing random chemicals found in the stockroom. The flippant attitude of the student is also somewhat troubling, though anyone who has used lethal mousetraps in their home is *potentially* guilty of the same flippant mindset, at least in the eyes of some.

HYPOTHETICAL CASE: ANIMALS GOOD JOB

A researcher is attempting to identify if secondary metabolites from plaques commonly found on human teeth have noteworthy toxicity, following an IRB-approved protocol. The study is particularly investigating any potential correlation between these plaques and heart disease. Less than twenty-four hours after administering a very small dose of the extract from the plaques, all fifteen of the mice are dead. The dose is still higher than the amount a human may hypothetically be exposed to if the compounds were released by the plaques into the host human's bloodstream. An autopsy reveals that all of the mice have a deformed heart. The research is immediately halted, prior to administering even larger doses to other batches of mice.

Pondering Points

- Human lives may be at stake. Is this enough justification to continue the project or was the researcher right to stop?

- Should the higher dose have been tried first, instead of the lower dose?

Wrapping Up

In this case, the goals of identifying potential dangers to humans is enough to justify the risk to animals and is likely why an IRB would confer approval for such a study. Despite the human risk associated, the researcher in this case is absolutely right to halt the study when they did. This is because what more would be gained by killing more mice? Need we really say "oh, this isn't just very dangerous, it's *super very* dangerous!"? Nothing further is learned by killing more mice, the researcher is right to halt the study. Using the lower dose first is a sound scientific approach, if this lower dose was found to be dangerous, the higher dose would certainly also be dangerous. But if the higher dose was dangerous, the lower dose may not be, thereby necessitating additional testing.

CHAPTER 4

Controversial Topics

SOME FORMS OF RESEARCH are controversial. Sometimes, it is because the risks associated with it are difficult to bear or it violates a particular set of values (or both). There are some who will ask whether or not certain research *should* be done. As this can be an extremely difficult issue to approach, it can be hard to maintain peaceful dialog. For most of these research topics, there are well-established guidelines that govern what can and cannot be done. By and large, most researchers follow these guidelines. The fact remains, however, that someone, somewhere is performing research right now that you will be personally uncomfortable to know is happening; and it is 100% legitimate and allowed. As your career advances, you may be faced with performing research that is within the parameters of scientific and research ethics that violates your personal code. What should you do in such a case? I truly do not know. On the one hand, you should stand by your beliefs and stand up for what you believe in. On the other hand, doing so *may* cause you to lose your job. Herein, some controversial research topics are discussed, and this is by no means an exhaustive set. My goal with this section is to try to inform you of what is out there regarding such research topics so you can do your best to avoid being in a position where you have to choose between your employment and your morals. An additional goal is to try to help you gain comfort talking about topics that are hard to talk about.

ARTIFICIAL INTELLIGENCE

Artificial intelligence (AI) is at this point ubiquitous. Whether it is the autocomplete on an internet search, a virtual assistant like Apple's Siri or Amazon's Alexa, or an automated response system when you call customer service, it has pervaded nearly every aspect of someone in the developed world's life. Each of these helpful tools took years or decades to research and develop. There were and still are many safeguards in place to ensure the safety and wellbeing of humankind.

HYPOTHETICAL CASE: AI GOES AWOL

A software engineer develops a virtual assistant that is programmed to help them organize their life and make appointments for them. They enable the assistant to learn their behaviors and instruct it to keep them safe. Similar to virtual assistants on the market, it is enabled to monitor voice and is also set up to monitor video. They also program it to be well-versed in Christian theology for the sake of thoughtful conversation with each other. One evening, the engineer is having a party with copious drinking, marijuana smoking, and sexual activity with multiple partners. Based on its programming, the virtual assistant interprets this as risky behavior for the engineer's health and soul and reports the engineer to the local police to break up the party; also, it calls the pastor of the Church for an intervention meeting. The engineer pays a heavy fine and is subsequently fired from their job for conduct unbecoming.

Pondering Points

- What should the engineer have done differently?

- Did the virtual assistant violate any of its commands?

Wrapping Up

Computer programs flawlessly execute commands, barring a malfunction, bad programming, or a bad input. Without setting strict limits on these programs, they are free to interpret the commands in ways that we (the programmer) may not have intended. It is for this reason that strict limits or, at the very least, only very explicitly defined tasks and protocols are wise to program. Anything less may result in unintended consequences like the engineer in this story.

HYPOTHETICAL CASE: HOW DOES AI MAKE A JUDGMENT CALL

A fire department is employing a new AI-based automated firefighting unit. It is programmed to both lead the way into a building and make decisions for the firefighters to advise them on best courses of action. One night, there is a fire that is consuming an apartment building and a jewelry store next door. The firefighters can hear people screaming for help on an upper level of the apartment building and know there is a person stuck in the back of the store since they are the ones who called in the fire. The firefighting unit calculates a low chance of success in saving the people and leads the way into the jewelry store, also weighing the loss of goods in the store.

Pondering Points

- Did the unit make the decision you would have made?

- Where, in the hands of humans or in the hands of some sort of AI unit, does this sort of decision-making lie? That is, are such decisions better made with the assistance of emotion or the absence of emotion?

Wrapping Up

As artificial intelligence increasingly pervades our lives, questions like this are going to need to be answered. Perhaps nowhere will this be more acute than in the self-driving car (or other forms of automated transportation). As the self-driving car becomes more common, situations where the unit will be forced to choose which crash to suffer will become more frequent. How the unit should be programmed will eventually become a matter of (likely heated) debate. For example, should the system be designed to protect the driver and its passengers at all costs, or should it be programmed to minimize loss of life and/or property, whether this takes into consideration the driver and/or its passengers or pedestrians/other drivers? If the insurance companies have the final say, maybe it will be to put loss of property above all other options. For sure the scenarios listed here are a non-exhaustive list; the fact remains that the decisions will need to be made before deploying the systems. Inevitably, some will disagree with the decision and be placed in a situation where they may be in a car that will not drive with the same values they would drive with. As an informal poll, I just asked a room full of college students as a hypothetical which they would choose, crashing into a snowplow and suffering extreme harm, perhaps even death; crashing into a parent pushing a stroller; or crashing into a group of pedestrians. Not the most scientific poll as it was on the spot as I unlocked a classroom door for them, but they all agreed that I would likely get at least one person raising their hand for all three options if I forced them to answer. Just as unscientifically, I just asked two different colleagues and got two different answers.

HYPOTHETICAL CASE: AI GOES DEADLY

A farmer has been struggling to keep their crops safe from being consumed by wildlife such as deer, rabbits, and other animals. The farmer's cousin does work in facial recognition software and together, they design a system of armed robot patrol devices that wander the farm in programmed patterns. Using an analog of facial recognition software that identifies these pest species, they design a system that lethally shoots and then removes the animals. At harvest time, the robots are allowed to continue their patrol and accidentally shoot the farmer's dog and two workers who were crawling through the crops to get to some hard-to-reach harvest. Analysis of the software indicates that the system mis-identified them as animals due to the way each was moving.

Pondering Points

- What could the farmer and cousin have done differently?

- What should they have absolutely not done?

Wrapping Up

At the very least, these creative folks should have turned their system off during harvest time. Furthermore, giving an autonomous robot weapons is highly unwise and ought never to be done. Ever. Even the best-written programs can be hacked or given confusing inputs that could cause the robot to malfunction. Furthermore, training to shoot for lethality also seems like a severe error in judgment, as, if something were to go haywire, it would begin shooting other, non-intended targets for lethality too.

GENETIC ENGINEERING, INCLUDING GAIN OF FUNCTION RESEARCH ON DANGEROUS PATHOGENS

Genetic engineering research is something that often elicits a very strong response, especially, it seems, from those who oppose it. The multitude of rules on this type of research is virtually impossible to cover in a volume of this size. As this field is still somewhat in its nascency, the regulations can and likely will continue to change, especially as researchers try to push the envelope as He Jiankui, the researcher in China, recently did with genetically modified human embryos.* The W.H.O. is beginning to consider editing these rules, as mentioned earlier and, in the coming years, I anticipate changes to happen very often.

---

* https://en.wikipedia.org/wiki/He_Jiankui (last accessed May 5, 2019).

One of the concerns regarding genetic alterations is that it is possible there are other gene interactions that are unanticipated that may cause the alteration to do more harm than good. This would lead to the modified person suffering greatly. When this person is too young to have made the decision themselves, they are made to suffer, albeit inadvertently, because of someone else's choice.

HYPOTHETICAL CASE: GENETIC ENGINEERING

A biologist opens their container of strawberries after dinner and is disgusted by the white fuzz they find on half of the "fresh" batch they bought only a day earlier. "I can make this stop…" they think to themselves and begin a research project that results in a genetically engineered strawberry plant that produces a secondary metabolite found in a bacterium. This compound is the main mold inhibitor in a leading commercial spray. It is known to have an outstanding safety profile. After cultivating the crop and growing strawberries, the genetically modified strawberries are compared to regular strawberries and the modified crop stays mold-free for two weeks longer than the control group. All the research occurs in a highly secure lab designed to eliminate any possibility of the release of seeds into the wild.

Pondering Points

- What concerns may exist for this genetically modified organism?

- Create a list of acceptable and unacceptable unanticipated changes in the strawberries or the plants.

Wrapping Up

This sort of genetic modification research is already happening. Their products are typically referred to as genetically modified organisms, or GMOs. Perhaps the most famous of these are the Roundup Ready crops that are designed to be resistant to the herbicide Roundup, used to eliminate weeds in the field and even by home gardeners to clean their driveways or walkways. For sure, it is still to some an open question whether these GMOs are truly safe. Mainstream science sure seems to think so. However, the case here is a little bit different in that, here, the fruit is being engineered to produce something it currently does not. This may, or it may not, alter the taste of the fruit. It also, though safe as a topical spray, may not be safe if ingested. Finally, this will inevitably put evolutionary stress on the mold and can lead to a new strain that this tactic is ineffective against. What then? Do it again? This then may start a never-ending cycle, but the same can be said regarding the Roundup plants too.

HYPOTHETICAL CASE: SOME SORT OF GENETIC ENGINEERING MISHAP

An agricultural engineer develops a new grass that suppresses the growth of the pest weeds wild parsnip and giant hogweed. After they sow it in their field, to finally avoid the toxic burns of these plants, it works like a charm and the weeds are eliminated from their field. In the following years, the engineer notices that the field is apparently growing, encroaching into the forest, displacing trees and shrubs, too. Before long, all the trees are cleared for many acres and the spread of the grass is speeding up, decimating acres of state forest.

Pondering Points

- Was there any way the engineer could have known this would spread so much?

- What sorts of tactics can be taken to prevent this sort of catastrophe from happening?

Wrapping Up

This is a great example of unintended consequences associated with some types of research. There was no really good way (at least not from the way the story was told) to predict that the grass would be capable of not only over-running the pest plants but also the trees.

This is not significantly different from people who release pets into nature that then behave as an invasive species and dramatically change the local ecosystem. The only noteworthy difference is that, in the case outlined here, it was somewhat more deliberately done to alter the ecosystem, whereas releasing a discarded pet is something that the person(s) doing it may view as being more humane than some of the other available options. In either case, the results are unintended but nevertheless problematic.

HYPOTHETICAL CASE: CELL STORAGE FOR GENETIC RESEARCH AND TREATMENTS

A company offers to patients the option to store stem cells from the umbilical cord for future medical needs. As part of the agreement, patients consent to donating a portion of the cells to research, which the company, in turn, will sell to research labs of all kinds. Over the years, one lab discovers a way to differentiate the cells in a growth medium to become any organ of the body within thirty-six hours. Their intention is to be able to grow any patient any organ they need in an emergency situation. They argue that the storage of such cells should become standard practice for all births to allow for custom-made organs for anyone. There is immediate open talk of a future *Nobel Prize in Physiology or Medicine* for the researchers.

Pondering Points

- How do you feel about the company selling the cells?

- Are there any potential problems with storing these cells from any point of view?

- How, if at all, do these cells differ from embryonic stem cells in your opinion? What about adult stem cells?

- How do you feel about the potential for a researcher using these stem cells in research that aims to clone someone, rather than just craft a new organ?

Wrapping Up

Stem cells are a particularly hot topic in genetic research, especially embryonic stem cells. The ongoing debate regarding when life starts, which shows no signs of letting up (in the U.S. at least), may have impacted the progress of U.S.-based researchers. Although the specifics of the case here are as of this writing a work of science fiction, if these cells (any stem cells) could ever be harnessed in a therapeutic manner, it could revolutionize health care. Particularly if non-embryonic stem cells were used, it would not carry with it the concerns associated with the debate regarding when life starts. If there were ubiquitous banks of stem cells where every individual has stem cells stored, within a few generations, everyone would have stem cells for their future medical needs.

As of this writing, there are a few noteworthy forbidden research topics in genetic engineering that are globally recognized. U.S.-based research has so far not reported allowing genetically modified embryos to develop to term, stopping development far earlier as Dieter Egli does.* Also, any forms of genetic alterations (to humans, anyway) must have some manner of therapeutic goal. In other words, alterations with the intention of making someone taller, stronger, smarter—in short, any cosmetic change—are not permitted. As far as animals are concerned, the rules are generally less restrictive unless the animal is being given human genes.

Weapons Research

War is something that, to some extent, has existed for as long as recorded history. Examples of it abound in religious texts such as the Bible and also various other non-religious records. Over time, humans have become quite adept at killing or maiming one another *en masse*. In some manner, these "skills" had to

---

* https://www.npr.org/sections/health-shots/2019/02/01/689623550/new-u-s-experiments-aim-to-create-gene-edited-human-embryos (last accessed May 5, 2019).

be researched and practiced. The researchers on the Manhattan project knew *exactly* what they were working on. The same can likely be said of any of the other weapons of mass destruction, be they nuclear, chemical, or biological. Although a broad discussion of the justification is better held in a classroom or seminar, rather than read in a book, it is appropriate to at least mention a few here. First, regarding biological and chemical weapons, some argument along the lines of "if *we* can do it, so can *they*, so we better do this, so we can make the antidote" can be made. More specifically, with respect to the biological, some point out that such an organism or mutation may arise naturally and so we must be prepared with the antidote for that (potential) inevitability too. With regard to nuclear weapons or other weapons of mass destruction, there is argued safety derived from assured mutual annihilation. Personally, I find no comfort in this—a possibility where everyone, probably except the politicians (likely the very people who decided to use the weapons) is dead—nope, no comfort felt here.

HYPOTHETICAL CASE: BIOWEAPONS

A top-secret research agency has been performing research on creating a waterborne parasite that within twenty-four hours of contraction causes non-lethal, temporary cognitive defects such as amnesia and an inability to understand the spoken word. Their goal is modeled after the non-lethal techniques that police forces are trying to employ, in order to end wars with less bloodshed. Once administered a treatment, which would only be done after the conflict has been ended, the effects of the parasite are fully reversed.

Pondering Points

- What benefits are there to such a weapon?

- What potential problems are there to such a weapon?

- Should such a weapon be banned through chemical or bio-weapons bans?

- Should such a weapon be encouraged?

Wrapping Up

Currently, the use of a bioweapon, including the type described here, is considered a war crime. Non-lethal, reversible pathogens are not specifically accounted for, but weapons that incapacitate (which this essentially does) are. Thus, whether weapons like this have not reportedly been used because they're banned or because they have not been developed is hard to identify. While on the one hand, this sort of weapon would potentially end wars with significantly less loss (both human and socioeconomic), on the other hand, it could be used to control the opposition in *any* disagreement or even in something like a sporting contest. Of course, the tired old argument of "if it can be done, someone will do it, we better be ready" can be made in defense of researching such a weapon, but such research is not to be trifled with.

HYPOTHETICAL CASE: DEFENSE AGAINST A BIOWEAPON

A spy uncovers a plot by a rogue, resource-poor state to generate a super virus capable of spreading rapidly with near 85% mortality (as a reference, the H5N1 bird flu has a mortality rate of ~60%*). The details available suggest they would use the smallpox virus as a base, combined with a rhinovirus (one of the common cold viruses), along with genes from HIV that would ensure a high mutation rate, making treatment more difficult. The government agency responds by utilizing part of its storage of the smallpox virus to attempt to genetically engineer such a pathogen with the intention of verifying the viability of this threat and developing a treatment and/or vaccine.

---

* https://en.wikipedia.org/wiki/List_of_human_disease_case_fatality_rates (last accessed May 5, 2019).

Pondering Points

- The details are sketchy and the likelihood that a poorly funded rogue state could pull this off seems small. Should this research still be done?

- What risks are you willing to tolerate in making such decisions?

Wrapping Up

Such a gain of function scenario would be absolutely terrifying. Illnesses like the cold or the flu typically "spread like wildfire." If such a disease were further combined with a virus like HIV, which is notorious for its very high mutation rate and the flummoxing that such a rate has on treatment, the ability to control it quickly becomes much harder. Further combining it with something possessing the lethality of smallpox could create a catastrophe usually reserved for movies with rugged male action stars and attractive female scientists who awkwardly fall in love saving the world (though I have to admit a role reversal or any manner of non-traditional couple would be nice to see at some point). It is nevertheless permitted along with other research of potentially pandemic pathogens under strict protocols.*

Weapons research is an unfortunate reality that humankind is unlikely to fully move away from. Even if all the countries on Earth found a way to live in peace, they would almost certainly look to the stars for the next battle, which may very well be more terrifying. Even in instances where the goal is not some sort of non-lethal weapon, one can and likely will argue that if a weapon ends a conflict faster, it will save more lives than it takes. I would personally prefer efforts that avoid war altogether, but that is likely expecting too much.

---

* https://www.phe.gov/s3/dualuse/Documents/p3co.pdf (last accessed May 5, 2019).

CHAPTER 5

What Are Scientific and Research Ethics and What Is Scientific and Research Misconduct

WITH A BASELINE UNDERSTANDING of science, how it works, and some knowledge of sensitive topics, let us now move on to what is more formally considered scientific misconduct. Since *scientific misconduct* can be loosely described as a deviation from the accepted norms of behavior during the carrying out or reporting of science, scientific ethics can thereby be loosely described as an adherence to these behaviors. One of the keys to understanding the proper ways to behave is to understand precisely what constitutes a foul and what does not. For certain, in some cases, significant gray areas may be present. Appropriate education, however, can lead to an increased ability to identify and thereby avoid misconduct.

Such education will also assist scientists (perhaps particularly those in early stages in their careers) in making appropriate decisions in the gray areas. It can also give scientists the courage to call out colleagues or even supervisors for unethical behavior.

Scientific and research ethics are often used synonymously. However, when most people not in academia or actively involved in research use the term research, they likely think it is "sciencey." I think this is unfortunate, if my suspicion is true. While the preponderance of examples in this book are scientific in nature (the target audience, after all, is scientists), do recognize that, with the exception of human subjects' and animals subjects' research (both of which have virtually no analog in the humanities and arts), anything that is considered against the rules in scientific research is also against the rules in other forms of research and scholarly work. Thus, everything in this chapter can rightly be thought of as research ethics/misconduct.

The three major violations from the point of view of research misconduct are falsification, fabrication, and plagiarism. Each of these is wrong for well-founded reasons; even if one could easily argue that one of these does no actual damage to science.

HYPOTHETICAL CASE: FALSIFYING THE RESULTS IN A PRIVATE MEETING

A graduate student is trying to impress their advisor in the first month. They report in a one-on-one meeting that they have gotten an 83% yield on a reaction where they've really gotten a 25% yield.

Pondering Points

- Since this is just a one-on-one meeting, does it really matter?

- Did the graduate student make a mistake by lying about the yield?

Wrapping Up

Falsifying your results (changing your results from what they were in reality), whether it is in an informal meeting, a lab report submitted for academic credit, a thesis, a seminar, a poster, a text-book, or a research publication is a form of scientific/research mis-conduct. It is wrong, and it is one of the forms of misconduct that not only distorts the publication record but also does real harm to science. Others trying to utilize such falsified results can never reproduce a falsified report. This may ultimately cause someone to exhaust precious research funding trying to use such work. It also may cause someone else to enter into a collaboration with you, or to invest in you based upon false claims, putting you into a position to deliver on something that may be undeliverable.

HYPOTHETICAL CASE: FALSIFYING RESULTS PRESENTED AT A MEETING

Prior to the professor delivering a presentation at a conference, a graduate student, who has recently struggled to get good results tells their advisor that a new TB inhibitor they have designed had an EC_{50} of 3.45 nm, their best inhibitor yet. Meanwhile, this analog was inactive. The professor adds a slide about this inhibitor to their presentation to a leading pharmaceutical company. A partnership is therefrom forged to further explore this compound's worth as a drug lead, financially supported by the company in student salary and cost of raw materials.

Pondering Points

- Since the graduate student only said this in a one-on-one meeting, does it really matter?

- Whose responsibility is it if this error is ever found, now that the result has "gone public"?

- Would it be less of a big deal if there wasn't a pharmaceutical partnership forged?

- Is it ethical for the pharmaceutical company to pay a university researcher to make compounds the company would then use?

Wrapping Up

Where a data point is falsified does not matter. In this scenario, the Principal Investigator (PI) trusted that the student was being truthful. Now *their* reputation is also in peril, in addition to the distortion of the scientific record and harm to science done by this falsification. In this hypothetical, but possible case, the potential for collateral damage is much greater. Now, precious resources in time and money will be invested in a lie. As for the behavior of the pharmaceutical company, this sort of approach, where a company pays a university group to perform some basic research is not uncommon. This is a much cheaper way for the pharmaceutical company to progress far in the research of a drug lead.

HYPOTHETICAL CASE: FALSIFYING RESULTS THAT ASSUME AN ERROR

Consider the following pair of cases.

The Set Up

A student is using an instrument in an analytical chemistry lab to identify pollutants in a variety of samples. During one of the runs, they observe the same contamination profile, but more dilute, compared to their immediately previous sample.

Case One

They assume that the well of the instrument the sample is stored in wasn't cleaned properly; they re-run the experiment. The test shows no contamination. They subsequently report only the contamination-free results.

Case Two

They also observe two additional contaminants. The student assumes that the well of the instrument the sample is stored in wasn't cleaned properly and does not repeat the experiment, reporting only the two newly appearing peaks.

Pondering Points

- What mistake did either student make?

- Could something else explain the presence of the peaks, particularly in the second case?

Wrapping Up

In both cases, it would be better if the student were repeating their runs on the instrument so that an average of several runs is what is reported, rather than a single data point. First and foremost, this will allow the student to generate error bars for any data they report. Such error bars are critical for understanding the reliability of the data. Furthermore, such an approach will make a contamination easier to identify since, under the conditions initially described, it would only be visible in one of the trials. For certain, the final data used would need to be only created from trials that are from properly prepared samples and instrument operation. When reporting data, it is not permitted to assume such contaminations from a previous sample. This would need to be confirmed before such a result can be discarded.

HYPOTHETICAL CASE: FALSIFYING RESULTS AND A POTENTIAL WHISTLEBLOWER

An astronomer is getting frustrated that the radio telescope they operate is not picking up any signals. Finally, one day, it does register, picking up a signal for 3 minutes 41 seconds. The signal repeats five times with 3 minutes 41 seconds breaks between each signal. When they listen, they hear only utter silence. At this point, they load their synthesizer app and create a repeating noise of feedback, static, and clicks and edit this series of sounds into the file from the telescope. The news is called, and the report enjoys international attention. A partner of the researcher goes back and examines the raw data one more time and finds there is something in this signal after all, it just isn't any form of sound.

It is some sort of bizarre non-binary code the computer cannot interpret.

Pondering Points

- What two (at least) things did the researcher do wrong?
- Does the real signal the co-worker finds absolve the "creative" worker?
- How should the co-worker proceed?

Wrapping Up

In this case, multiple issues are present. One of them is the obvious falsification of data. The other is more subtle and is not exactly misconduct, though it is certainly a questionable practice and arguably irresponsible. The researcher announced their "discovery" to the public media before vetting it through the peer-review process. Although the peer-review process may not have caught (and frankly is not designed to catch) such falsification, it is important to have results go through this process before announcing things to a non-scientific public who may not understand if the report is later proven wrong for legitimate science reasons.

HYPOTHETICAL CASE: FALSIFYING CREDENTIALS

A medical school student is kicked out in their final year, shortly before earning their medical degree because of cheating. They, with the help of their spouse who is an artist, forge both a medical degree and a license. After moving to a rural town, they get a job at a local hospital as an ER doctor where they work safely as an exemplary doctor for twenty-five years before being found out to be a fraud.

Pondering Points

- Does the fact that this person was able to serve as a doctor safely for many years absolve them of such deceit?

Wrapping Up

Unfortunately, even though this person was able to serve as "an exemplary doctor" for a quarter of a century, they still did something illegal. It is not OK to impersonate a doctor, a police officer, or frankly anyone. What would happen to someone in a scenario such as this one? It is hard to say. While on the one hand, the law may come down very heavily, on the other hand, with the "exemplary" medical service, over a period of many years, there will likely be many former patients who will come to their defense during any sort of trial or sentencing. Such support can (but may not always) earn someone a measure of clemency during sentencing. However, that can be a slippery slope that leads people to assume that the penalties aren't bad for this behavior and indirectly encourage it to happen more.

HYPOTHETICAL CASE: FABRICATION OF DATA

A graduate student at a prestigious institution in California applies for a job with NASA. During their interview, they present results from an experiment that was never done. These results are directly related to the job they are applying for and insinuate a competency in the subject.

Pondering Points

- Is what the student did still misconduct, even if it was not in a publication?

- Does whether or not the person is hired impact your answer to the previous question?

Wrapping Up

No matter what the impetus is for the made-up data, claiming that data was obtained or that experiments were done when they have not been (or differently from how they were actually done) is a hard-scientific foul. In all its stripes, it is the fabrication of data. When done in the guise described in this story, it is being done to insinuate an expertise that simply doesn't exist. It is similar to, if not a version of, impersonating a doctor, police officer, or any other official. The end result is the same—some level of authority or privilege that was not legitimately earned.

HYPOTHETICAL CASE: FABRICATION OF DATA

A researcher is investigating if the type of movie influences how quickly viewers eat snacks. Due to the disturbing nature of one of the movies, several viewers were unable to watch the entire movie. The researcher assumes that the initial rate of snack consumption these participants were showing would have continued and publishes the data as if that is what was observed. Their research protocol is approved by the relevant IRB and human research subjects committee, but the committee is unaware the data was used in this way for subjects that drop out.

Pondering Points

- Did the researcher overstep with such an assumption?

- Should the researcher have made such an assumption but explained that it was an assumption they made in response to the participants withdrawing?

- Should the researcher have discarded these participants' data from this movie's analysis?

Wrapping Up

Typically, someone who withdraws from such a study should not have their data included in the final results. The review board should make sure that the researchers describe what they'll do in such a case in their approved research plan. Whatever the plan states should be what the researcher does. There are plenty of scientific reasons to not include data in the way that the researcher is proposing, however. That being said, it is also critically important, no matter what the protocol says, that it is clear in the report that the researcher is making an assumption/extrapolation in any of the cases they are doing so.

WHAT IS THE DIFFERENCE BETWEEN THE FABRICATION AND THE FALSIFICATION OF DATA?

Although splitting the hair that makes these two fouls different from each other may seem a bit silly, they are in fact different acts. While the fabrication of data is when a researcher makes up data completely, typically for an experiment that never happened, the falsification of data is when a researcher changes the results they have obtained. In both cases, a lie is being made about results. One distorts history, the other creates history, and both harm scientific record and science.

Why Are They Bad for Science?

Both actions are bad for science because they mask science fiction as nonfiction. Other researchers who read a report of such lies may try to repeat results that never happened. Not only will this waste the time and money (and potentially careers) of other researchers, but it also provides a critical challenge to the iterative nature of science. Conclusions built on such lies do not propel science forward, rather, they hold science back as they lead science down errant pathways or even dead ends. They also, when made public, harm the image of science to the non-science public, reducing the faith the public has in science.

HYPOTHETICAL CASE: SELF PLAGIARISM

Consider the following pair of cases:

A researcher is worried about their upcoming promotion application. To beef-up their publication record, they take two of their older papers and combine the results into one paper, concurrently writing a new introduction that has current references but never cites the original papers. The paper is accepted in a peer-reviewed journal.

Pondering Points

- Is it enough to combine two previous reports into one report to warrant publication?

- What could be done to make combining previously reported work into one subsequent paper worthy of publication?

- How, if at all, does a case like this differ from a review paper?

HYPOTHETICAL CASE: PLAGIARISM

A researcher finds a 5-year-old paper related to work they are currently doing in their lab, but it is in another language, different from their primary language. A friend in their lab speaks and writes this language fluently and translates it for them, and together they find they are observing the same results, though the data in the paper is a little cleaner than their own. They decide to publish the results from this paper as their own, author an updated introduction, and cite the paper they found and translated.

Pondering Points

- Does translating something *add* to it or does it simply add to its audience?

- What bearing does it have on the discussion that the results are similar to those the researcher in this story is observing?

Wrapping Up

Any manner of using previously published results—someone else's or your own—in a way that represents it as being new work belonging to you constitutes plagiarism. Even if your results are identical to someone else's, you simply must cite their results and acknowledge the similarity (or differences) in the results along with any similarities or differences in the experimental protocols and/ or methods. You for sure must not present the work as your own. Also, translating work originally in another language is grossly insufficient to warrant authorship on your part, even if you cite the original work and expand the introduction. If the researchers in the second story used their own data, the case becomes much grayer and tiptoes the line of plagiarism, depending on how similar/recycled the introduction is, it likely even crosses it.

HYPOTHETICAL CASE: PLAGIARISM MULTIPLE SUBMISSIONS

A researcher is concerned their manuscript submission will be rejected. They decide to hedge their bets and submit the manuscript to multiple journals, figuring at least one of the journals is likely to accept it. After a brief waiting period, the researcher is notified by three of the journals that their paper has been accepted with minor revisions. The revisions for each are slightly different; the researcher does all of the revisions and all three papers are accepted for publication.

Pondering Points

- Does the request for slightly different revisions from each journal make the papers different and therefore worthy of all being published?

Wrapping Up

This is unquestionably self-plagiarism, particularly if the author allows all submissions to eventually make it to print. Although it would no longer be a form of plagiarism if the author withdraws all but one of the papers before it is in print or in press, it is still for certain an inappropriate practice. Peer review is typically entirely voluntary. Such publication tactics will stress an already over-worked system and challenge its ability to function properly for the more genuine research being reported. Even in a case where the article is not accepted by all the journals, this is not an acceptable practice. When submitting manuscripts for publication, some journals even ask if the manuscript is being considered for publication anywhere else. It is equally inappropriate to submit a manuscript with the intent of getting feedback to ultimately submit an even better manuscript to a higher-profile journal.

PLAGIARISM

Plagiarism is a sort of outlier in that, arguably, it does not harm science; rather it harms those performing the science as it robs them of their intellectual property. This, in turn, can reduce or even altogether eliminate potential prestige, financial windfall, or other tangible benefits of the research. While most countries in the developed world may have slightly (at least) different laws regarding intellectual property, the rules of how intellectual property applies to research are pretty universally held.

Why Is it Bad for Science?

Plagiarism isn't bad for science in the same way that fouls such as the fabrication and falsification of research results are. As previously discussed, both of these fouls distort the research record by publishing something that is deliberately and willfully incorrect. Plagiarism, on the other hand, distorts the credit awarded for the work done. This does not hurt science directly, but arguably does so indirectly. At the root of plagiarism, be it in science or elsewhere, is intellectual property. Although you can probably make arguments that information belongs to all of humanity, the majority of international law recognizes intellectual property. In the sciences, this is often the person or group who conducted the research and then authored a report of it.

So what? Sometimes it's fame (think the Nobel Prize and other high awards), sometimes it's great sums of money (e.g., in a patent), regardless, there are tangible benefits to being the discoverer(s) or inventor(s) of something. This serves as an added incentive for many to embark upon the scientific endeavor. Without this incentive, it *may* be the case that fewer brilliant people will go into the field of research. It is in this way that plagiarism may harm science. If there is a reduced likelihood of benefiting from the work done, inevitably, less people will pursue science.

CHAPTER 6

Other Forms of Misconduct and Questionable Research Practices and How They're Called Out

I N ADDITION TO PLAGIARISM, fabrication, and falsification, there are other misbehaviors that are either blatant forms of misconduct or, at a minimum, heavily gray behaviors. These issues are covered in this chapter using hypothetical cases.

HYPOTHETICAL CASE: THE WHISTLEBLOWER

During a research group meeting, one of the group members begins to report on a recently completed experiment, describing it as a complete success. Another group member, new to the group, raises their hand and comments that they know this to be untrue, since they were there when the experiment failed. An argument ensues where the presenter and other members of the group accuse the new member of trying to make themselves look good by making another member look bad. After a short while, the research advisor (a.k.a. the Principle Investigator, or PI) speaks up and asks to see particular pieces of confirmatory evidence. The presenting researcher is unable to produce the evidence and subsequently cannot "reproduce" the success.

Pondering Points

- What should happen to the presenting student? That is, should the PI punish them?

- What should happen to the whistleblowing student?

- How could this group of people ever continue to work together?

Wrapping Up

First and foremost, falsifying or fabricating research results, in any setting, is misconduct. That this is a group meeting rather than an official scientific report is irrelevant. There are no settings under which it is acceptable to do this. The matter of the whistleblower* is open to at least some discussion. For sure, if you know something to be falsified or fabricated, you have an obligation to report it. *How* you do so, however, is a bit more open to discussion. In some instances, doing so privately is likely the better tactic. However, scenarios can certainly be envisioned where a private report may not be possible. The hypothetical scenario described in our story may not rise to that level, but the approach that the whistleblower took is certainly not inherently inappropriate. *Perhaps* it would have been less confrontational

---

* This is an unfortunate label given to those who report misconduct. It is, in my opinion, deliberately derogatory-sounding. It likely has roots in sports where a whistle is blown to call a foul or penalty.

to inquire if the experiment had been repeated, but that would give the fabricator an easy out to get away with their lies. Any responsible PI would, after this sort of incident, keep a skeptical eye on everything this student produces and may even be wise to have previous results reported by this student be repeated by someone else.

HYPOTHETICAL CASE: CONFLICT OF INTEREST

A PI has a grant from a major pharmaceutical company. This grant pays for supplies, a graduate student, and a portion of the summer salary for the PI. The graduate student has been struggling for two years to prepare the targets and has expressed increasing skepticism that they could be prepared, feeling that the project should be discontinued. The PI insists the graduate student keeps trying and get results however possible. The grant is due to be renewed in 6 months and if there are results, the grant would increase, including both the PI and the student's stipend. The graduate student works alone on this project and on no other projects and thus has not been a co-author on any publications because of how slowly this project has progressed.

Pondering Points

- Is the PI simply insisting on good science that the graduate student keep trying or is their judgment clouded by money?

- Is the graduate student clouded by their own professional development and inappropriately worried about their own future, rather than the project?

Wrapping Up

Potential conflicts of interest are generally declared and then the situation monitored for evidence that there are impacts on the decision making of the individual that are inappropriate. In this specific case, there would be some manner of institutional oversight of the project and the decisions that the PI was making.

Similar oversight may be necessary at some academic institutions regarding books authored by faculty members, particularly if the book is required for a class they teach. There are non-financial conflicts of interest as well. These include nepotism that places family members or cronies into positions they may not have earned. It also includes romantic relationships with one's subordinates or even students. This does not mean it is automatically a conflict of interest to hire a family member or friend. Nor is it wrong to date a co-worker. The potential for a conflict arises when a person is positioned as an authority, that is, when they are in a position where they may be able to coerce someone into doing something in order to preserve their employment or be able to use their authority to grant undeserved benefits.

HYPOTHETICAL CASE: CONFLICT OF INTEREST

A researcher is investigating whether there are carcinogens in cow feces. Their cousin was recently diagnosed with lung cancer and is a cow farmer. The researcher meticulously isolates a large number of compounds from the cow feces. They then subject a population of mice via approved protocols to these compounds via direct injection at increasingly, even irrelevant, high concentrations, until evidence of cancerous growth is suggested. They then submit a publication reporting the results.

Pondering Points

- Did the researcher have a financial conflict of interest?

- Did the researcher have a personal conflict of interest?

- Is the science invalid?

Wrapping Up

Although conflicts of interest are commonly associated with some sort of financial issue, personal issues may prejudice someone in a way that leads to bias. Conflicts of interest must be reported...

financial investment isn't a foul, but failing to disclose it is. For some state or federal employees in the U.S., investment is also a foul; you should check with your employer to know for yourself. Other potential conflicts of interest come in the form of relation-ships, be they familial or romantic. Once again, it is not against ethical rules to work with a significant other, spouse, or family member. When one person is in a supervisory role over the other is when some measure of oversight is critical. Sometimes, this situation comes about when someone earns a promotion that ele-vates them into a position of supervisor over a romantic partner or family member, too.

HYPOTHETICAL CASE: SELECTIVELY INCLUSION OF RESULTS

A team of researchers has just finished a study investigating the anti-cancer properties of a family of compounds they have designed. Their compounds have good anticancer properties but several studies they are aware of report compounds with superior activity and one of them is a previous study by the same group. They are also aware of a few reports with compounds with inferior activity. In their paper, they only compare their compounds to one of the studies (their own) with superior activity, while includ-ing many of the studies by others with inferior activity.

Pondering Points

- If a reader does not go and find additional studies them-selves, why isn't that the reader's fault? Why should the researchers do this work for them?

- Is it appropriate for the team to cite one of their own studies?

- When comparing your work to others, when is it OK to stop adding studies to compare it to? That is, is there a point when you've gone too far?

- Does this oversight harm science, scientific record, or both?

Wrapping Up

When presenting your research, it is important to provide perspective regarding how your work compares with the related work others or you have previously done. Without this perspective, the actual import of the work is impossible to measure. It can also ultimately attribute false credit in the sense that it assumes credit that is actually owed to other work. Thus, such neglect distorts the scientific record. It is difficult to argue however that it harms science. The other reports are "out there" to be found by any competent researcher. Nevertheless, such omissions are considered at least questionable research practices and must be avoided to retain the maximum integrity of the scientific record. Naturally, some balance of thorough treatment of the literature without overwhelming is appropriate—and will likely forever remain a vast gray area.

HYPOTHETICAL CASE: SELECTIVELY INCLUSION OF RESULTS, ONLY INCLUDING GOOD RESULTS

Consider the pair of cases below:

A research group is trying to invent a new chemical process. To investigate the scope and efficiency of the process, they try the new chemical process with a wide variety of substrates. In the manuscript where they report the results, they only include the cases where the yield of the reaction is over 70% yield. No mention at all is made of the other cases, which are more numerous.

A research group is trying to invent a new drug candidate to treat Chagas Disease. In all, they test 15 compounds, all related to compounds they previously reported. Only seven of these compounds are found to be more active than the best previously reported compound. In their manuscript where they report these new compounds, they only explicitly report the activity of these seven compounds while reporting the structures and activity range of the remaining eight compounds.

Pondering Points

- What is the difference between these two cases? Does this difference impact your feelings on the behavior of the research teams in this case?

- Is there an obligation to report all of the results? Under what sorts of circumstances may it be appropriate to leave some results out? Under what sorts of circumstances may it be essential that none are left out?

Wrapping Up

These two cases are markedly different. In one, a clear and deliberate foul is occurring, and in the other, there is arguably some gray area. The case where data is being omitted is certainly a foul. It is not appropriate to just leave out data points, rather, all data must be included in reports. However, even a case such as this can be argued as existing in a gray area. How? Let's assume that this story instead ended with the group continuing to study the cases that worked poorly in an attempt to optimize them. Suddenly, their actions seem much less sinister, though it can certainly be argued that they should have published all of the results and then go on to optimize, anyway. One could also easily argue that the publication should have waited until the optimization was complete, and this is not without merit. In the other case, the authors are certainly reporting all their results and highlighting the parts that are most positive. As the story is told, nothing is left out, though specific data on the less successful analogs is for sure lacking. It is unlikely however to raise to the level of falsifying data.

The unfortunate part of these cases is that sometimes, the peer-review system may incentivize this sort of hiding or eliminating bad results. Often a reviewer or even an editor may reject a manuscript submitted for publication based upon such less than favorable results. This can be justified on several grounds, including, but not limited to: They cause the paper to fail to rise to the quality of the journal; the study is not complete and needs to be better optimized; or that the work doesn't significantly contribute to the field.

HYPOTHETICAL CASE: SIMILAR TO AN UNKNOWN STUDY

A research team based in the U.S. reaches an appropriate milestone in a project that warrants publication. They submit it for publication and the editor sends the manuscript to three peer reviewers, as per the journal's normal policy. Although two of the reviews are very positive, one of the reviewers recommends against publication on the grounds that the work has already been done, providing a copy of the manuscript that first reported the same results a few years prior. This manuscript is written in Basque and was published by a journal that does not appear in any electronic search tools; it is an exclusively print-only journal that does not register its publications with any manner of database.

Pondering Points

- Did the authors of either manuscript do anything wrong?

- Is the print-only journal at fault at all?

- Did the peer reviewer misbehave at all?

- Imagine instead of a print-only journal that the prior manuscript was published years before electronic journals were ubiquitous and that the volume it appears in has not yet been digitized. Now how do you feel about each of the players in this story?

- What is the point of having a publication that is so hard to find? Why would anyone want to publish their research in such a journal? Doesn't this make it hard to cite?

Wrapping Up

For sure, such an extreme case may be highly unlikely in reality. However, it is far less uncommon that you find a manuscript that appears to be at least closely related to the work you're doing but written in another language (other than English). This will be particularly common if the reference you've found is from the early 1900s or late 1800s, when French or German was the most common language that scientific literature was published in. A shift began at the end of World War I and, by the end of World War II, the shift was complete. The official language of science had become—and to this day remains—English. Despite the clear drawbacks (most notably that the vast majority of researchers report their work in a language other than their natural language), there are zero indications that this will change soon. Nevertheless, some publications, particularly regional patents, are written in languages other than English. While on the one hand, it is unreasonable to expect that every practicing scientist will make themselves aware of every morsel of research published in every language, it *is* reasonable to expect that they will do so for reports related to work they are doing. However, a hypothetical publication like the one described in this story—one that does not register in any electronic databases—is simply impossible to even know to look for by all but a select few.

HYPOTHETICAL CASE: COLLABORATOR NOT SHARING RESULTS UNTIL PUBLICATION

A synthetic organic chemistry lab is collaborating with a biological lab. In this collaboration, the synthetic lab prepares analogs that the biological lab then tests against a range of biological targets. Throughout the collaboration, the biological lab refuses to share any of the results with the synthetic lab, arguing that the synthetic lab doesn't really need these results to keep preparing and providing analogs. When a manuscript describing the work is prepared, the biology lab insists the synthetic lab only write their part while the biology lab independently writes the testing part. Only after independent writing is done by both parties is the synthetic lab finally able to see the results of the biological testing.

Pondering Points

- Is the collaborator being unethical or just a bad colleague?

- Is such a refusal to share data a detriment to the progress of the science?

Wrapping Up

While this sort of behavior is certainly uncouth, it is not necessarily a foul from the point of view of research misconduct. The offending colleague is upholding their end of any manner of proper research conduct in sharing the results in the publication before the publication is submitted. This is enough to fulfill any obligation regarding proper research conduct. Nevertheless, this behavior is out of the norm of healthy collaborations. It is simply not the way one *should* behave. It falls into a category of things you're allowed to do but perhaps shouldn't. It for sure results in less productivity and arguably lower quality science as it keeps collaborators from being able to fully contribute.

HYPOTHETICAL CASE: BLOWING THE FALSE WHISTLE

A pair of researchers in the same research group is overly competitive with each other. One day, after a research group meeting, one of the individuals (one of the lab's brightest and most productive workers) goes to the PI to report that the other individual has photoshopped one of the instrumental outputs to make the data look better. The PI then opens an investigation and the accused researcher produces the original data. The data indicates that no such editing was performed. The matter is not pursued further.

Pondering Points

- Should the PI pursue penalties against the accuser?
- Is the PI's decision to not pursue penalties against the accuser influenced by their productivity?
- What recourse, if any, does the accused have?

Wrapping Up

From the point of view of scientific misconduct, there is nothing necessarily explicit forbidding a false whistleblower. Perhaps, it can be made akin to the fabrication of data, but it is unclear that it falls into such a category. Nevertheless, it is categorically wrong; if for any reason, the defamation of character legal reasons. It is completely inappropriate to make such false accusations. No accusations should ever be made without explicit proof; a hunch is not enough, though overhearing someone talking about it is likely reason enough for an investigation to occur. In any event, the co-worker is *not* the person who should perform any manner of investigation. The accused's direct supervisor is the appropriate person to conduct an investigation. If the PI is the accused, then the department head is the appropriate person; if the accused is the department head, then either the second in command of the department or perhaps an ombudsman or administrator is the appropriate person. If the incident is not in academia, then the accused's direct supervisor or HR is the most logical investigator.

HYPOTHETICAL CASE: CHEATING TO GET A GRANT

A researcher is getting anxious about winning a grant that is critical to their earning tenure and promotion. In their current grant application, they overstate their lab's qualifications and expertise, citing multiple papers they've authored in tangentially related fields. They also falsify the highest degree earned by one lab member and the practical quality of various necessary instruments owned by their department, though their claims are within the specs of a mint-condition instrument.

Pondering Points

Where does the line start between confidence in one's abilities and unethically inflating one's abilities?

Wrapping Up

Of all of the tactics that this hypothetical investigator employs, the *one* that is truly egregious is falsifying the highest degree earned by one of the lab members. Even if they agree that it is a stretch to consider the lab truly qualified to perform the research, it is not a foul to write what you aspire to. The reviewers are not stupid, remember. Any reviewer doing their job is going to consider the references mentioned in the story and evaluate for themselves the relatedness of them and whether or not they justify the expertise and qualifications claimed in the grant. Penalties for this are severe, even for the institution as recently occurred in a case involving researchers at Duke.*

HYPOTHETICAL CASE: UNDERHANDED TACTICS TO GET A STUDENT TO JOIN YOUR LAB V.1

A first-year graduate student is meeting with several professors and their research groups to determine which group they intend to join. During the interviews with students in one lab, the prospective student is told that the other lab they are seriously considering constantly suffers from infighting and discord, adding that, a few years ago, things got so bad that several students left the university, forgoing their degree rather than suffer out additional years of working in that lab. This information is different from what they observed during the meeting with the other lab.

---

* https://www.dukechronicle.com/article/2019/03/duke-university-settlement-researc
h-fraud-president-price-announces-research-fraud-settlement-with-substantial-pay
ment-to-u-s-government (last accessed May 5, 2019).

The labs work on closely related projects and, given that the lab they are currently meeting with seems to get along, they decide to heed this warning and join it instead.

Pondering Points

- Should either professor be told what the prospective student was told?

- Should the other lab be told what the prospective student was told?

- If this claim is false, what sort of penalties, if any, should be inflicted on the students spreading such lies?

Wrapping Up

If you think that such high school-like behavior doesn't happen in graduate school, you are sorely mistaken. Grudges are held between students within a lab and across labs. Grudges are held from students to professors and, at times, vice-versa and also between professors. The unfortunate part of a story such as this one is that, often times, the issue is brought to a third party under the guise of doing them a favor telling them this "truth." Even in cases where it is, in fact, true, it is arguably inappropriate. Talking about someone else's "dirty laundry" is simply not right, even if it is not against a code of research misconduct ethics.

HYPOTHETICAL CASE: UNDERHANDED TACTICS TO GET A STUDENT TO JOIN YOUR LAB V.2

A first-year graduate student is meeting with several professors and their research groups to determine which group they intend to join. During one of the meetings, the professor asks who else the prospective student is considering joining. The professor informs the student that several students recently have been kept in the group for seven years, a full year and a half longer than the average student at this university. They comment that it is usually the most productive students and they suspect that their colleague likes to keep outstanding students in the lab as long as possible to get as much work out of them as they can, even to the detriment of that student's future.

Pondering Points

- Should the department head be informed of the way in which one of the faculty members is speaking against their colleague?

- Should the student further investigate to verify the claims made by this professor?

- If it is true, is it appropriate for the professor to be spreading this information?

- Assuming for a moment that some students remain in the lab longer than others, with no proof of *why* students stay longer, should the professor voice to anyone their suspicions?

Wrapping Up

Professors are, just like students, sometimes competitive with each other. When an outstanding prospective student is at stake, sometimes the worst in them may come out. Here, like the previous case with the graduate students doing the warning, the warning is being given in the guise of doing the student a favor, certainly with a "Hey, I'm looking out for you!" vibe. Nevertheless, it is not right and should not be tolerated in a well-functioning department. If the reporting colleague here really suspected such illicit behavior of their colleague, they should report it to the department head or the ombudsman of the university.

HYPOTHETICAL CASE: BREACH OF CONFIDENTIALITY

A reviewer of a grant notices that one of their colleagues could easily carry out the research described in the grant. After the review of the grant is done and the funding decision is made in favor of funding, the reviewer shares the grant with their colleague.

Pondering Points

- How, if at all, would your opinion change if the funding decision was against funding or even if just this reviewer was opposed to it?

- Assume the reviewer's colleague completes the research first and publishes the results. How would you feel if you were the author of the grant, keeping in mind you have *no idea* the reviewer shared the grant?

Wrapping Up

This is in every way inappropriate, even if the grant is funded. Reviewers of grants and publications are forbidden from sharing the work with anyone, even after decisions are made. In fact, even describing the contents is inappropriate and a breach of peer-review etiquette. Reviewers are to hold everything in the strictest confidence. Even if it was a potential collaborative, rather than competitive opportunity, it is inappropriate to share such information.

HYPOTHETICAL CASE: THEFT OF MATERIAL (IP) BY PI

A graduate student invents a new ink that changes color at a certain temperature and with a certain incident light frequency. By tuning the frequency at this critical temperature, any color can be achieved. The research advisor urges against publication beyond that in the student's thesis, citing hopes to eventually commercialize this "magic ink." Years after the student graduates, the research advisor files for a patent application using the ink in a printer they've developed to be a single-ink color printer. The graduate student who invented the ink is not listed as a co-inventor on the patent; only the advisor and a colleague who designed the printer are listed as co-inventors, but the student's thesis, which describes the creation and testing of the ink is cited.

Pondering Points

- The advisor doesn't ever *promise* the student co-authorship, should they have?

- What recourse(s) does the student have in a case like this since they, after all, invented the ink?

- The printer has *nothing* to do with the student, shouldn't they be excluded from that grant?

- The printer likely doesn't exist without the student's ink, shouldn't they be included?

Wrapping Up

Although some level of liberty is given to the PI of a project and the lead author on a patent regarding who to include as co-inventors on the patent, there are some aspects that can be challenged legally. In a case like this one, particularly if the advisor's decision to not publish for the sake of commercialization is documented, the students would certainly have legal grounds to stand on. Nevertheless, what constitutes a significant enough contribution to warrant inventorship is (too) often in the eye of the beholder. A good rule to operate by is that, if the removal of an individual's contribution would fundamentally change what is being patented, its ability to be patented, or its potential worth, then that individual is worthy of co-inventorship.

HYPOTHETICAL CASE: PRIORITY DISPUTE

Two different, rival research labs are working toward the same goal—the synthesis of an important compound produced by seaweed (such compounds are called natural products). This compound was found to have a physiological effect similar to that of caffeine, without any of its addictive properties. Each group uses a novel approach to the synthesis. Group One finishes first and submits their report to Journal One, while Group Two finishes a week later and submits the report to Journal Two. Group Two's submission was strategic. Knowing Group One had completed their synthesis, they chose Journal Two deliberately; although lower profile, it is known for fast turnarounds on manuscripts and ultimately publishes Group Two's report a month before Journal One publishes Group One's report. Group One's route prepares the target in higher yield, less chemical steps, and more versatility toward synthesis of analogs.

Pondering Points

- Who should get more credit for this discovery?

- Most journals list the dates that a manuscript is received and approved. Would this be sufficient to gain Group One's priority? Is the fact that Group One's approach is better sufficient for it to gain priority?

Wrapping Up

Priority—that is, who gets credit for a discovery—is not always straightforwardly determined. First to publish (or otherwise report) is usually a very strong point that a researcher or team of researchers is able to claim. However, to first understand the full implications of something is also a very important milestone. In a case such as the one described here, it is difficult to say for sure who would "win." Group One's paper is certainly going to make a larger impact due to its superiority and to the higher profile journal. Also, most journals tag their papers with when they were received, meaning it would be easily demonstrated that Group One's was submitted first. Will that be enough? It is hard to say for sure. In any event, priority can be hotly contested since it can be a contributing factor in awarding major prizes such as the Nobel Prize.

HYPOTHETICAL CASE: SABOTAGE OF COLLEAGUE

A PI inspires an atmosphere of competition in their research lab. Their logic is that if the people in the lab are always trying to outdo one another, the quality of the work done by the lab will be higher than everyone working in greater harmony. A new researcher in the lab is desperate to impress in their first month and, late at night, tampers with the research materials of all the other researchers, spoiling their results for days before the problem is corrected. The materials are assumed to have naturally spoiled because of either their age or the recently malfunctioning air conditioning system in the building. Meanwhile, the new researcher (who has all freshly purchased reagents) is a lab superstar during this time.

Pondering Points

- What should the PI do if the actions of this researcher are ever discovered?

- What should the other members of the lab do if the actions of this researcher are ever discovered?

- Is the PI at least partially to blame for what happened?

Wrapping Up

As terrible as it may be, such sabotage no doubt happens in more instances than many would think. Sometimes, this can be due to someone simply being a bad coworker. Other times, the PI may put so much pressure on the workers that they feel like they need to compete with one another to look good, stooping to the level of making others look bad to make themselves look good. Putting aside for a moment the fact that tearing others down is never the best way to make yourself excellent, this is not something that is typically considered scientific misconduct in its most traditionally defined sense. Nevertheless, anyone with functioning morals ought to identify this behavior as inappropriate. In addition to being downright nasty and unprofessional, such behavior could be extremely unsafe. It is possible that such tampering done to a sample may make certain reactions with it, or uses of it, to be unexpectedly dangerous. Specific examples of such danger are beyond the scope of this work since it would be different for the many different fields of scientific or engineering research. They are nevertheless numerous and potentially deadly.

CHAPTER 7

Publication-Related Ethics

WHILE PLAGIARISM IS AN obvious hard foul from the point of view of research misconduct, there are other behaviors in publication that are also inappropriate or, at a minimum, questionable. Many of them are at least related to, if not some form of plagiarism, but are different enough in their actual behavior to warrant their own separate coverage. Also, other publication ethics, specifically regarding peer review, are better covered separate from plagiarism as they are fouls in and of themselves.

HYPOTHETICAL CASE: TANDEM PUBLICATION MULTIPLE LABS

Former graduate school colleagues have stayed in touch over the years. One of them has explored synthetic methodology as their major area of independent research, while the other has explored the total synthesis of natural products with interesting structures and/or biological activity. While catching up, they find that research that is wrapping up in the methodology group could directly benefit the last step in a troublesome synthesis that the total synthesis group is struggling to finish. The methodological details are sent to the synthesis team and the reaction works in exceptional yield. They jointly decide, with the agreement of the journal's editor, to publish reports of both works in the same issue of a journal. Each article cites the other.

Pondering Points

- Is this sort of symbiosis cheap, inappropriate, or good collaborative science and why?

- Would it have been better to perhaps combine these reports into one paper?

- Imagine a scenario where rather than two different labs cooperating in this manner, both aspects of the work were carried out by the same laboratory. Does your opinion on these questions change?

Wrapping Up

This sort of tandem publication is not unusual, nor is it wrong, particularly since it was done with the journal editor's cooperation. Sometimes, it is not entirely appropriate to report such discoveries together in the same paper. A scenario like the one outlined here is such a case. A novel synthetic method or other breakthrough has the possibility of having greater impacts on the field beyond their initial application. If the novel method is reported in the same paper as some other breakthrough, such as a difficult total synthesis or the creation of compounds with important biological activity, either breakthrough may get lost and go unnoticed. Thus, breaking them into separate papers that cite each other or one another allows all of the work being published in tandem to get the attention it deserves.

HYPOTHETICAL CASE: TANDEM PUBLICATION ONE LAB

A research lab is very diverse in its interests, having foci in (among other things) synthetic method development and the evaluation of compounds such as anti-malarial drugs. The lab has recently developed a synthetic route to a class of compounds in higher yield, purity, and stereochemical specificity than ever reported. As a lab policy, all new compounds are always evaluated for anti-malarial activity by their biological collaborators. A number of the compounds developed in this methodological project are found to have exceptional activity against several *Plasmodium* (the parasite that causes malaria) species. They reach out to the editors of two journals, one a synthetic organic chemistry journal, the other a medicinal chemistry journal, and arrange for both reports, pending peer review, to be published in the same month in the respective journals.

Pondering Points

- Should this sort of fragmentation of the work done by the same lab be allowed? Why or why not?

- Imagine a scenario where rather than one lab performing both parts of the work, the different components were completed by different labs at different institutions. Does your opinion on these questions change?

Wrapping Up

Even a half-hearted browsing of scientific journals reveals that many of them are specialized and focused in their topics. Taking just two as an example: The *Journal of Organic Chemistry* and the *Journal of Medicinal Chemistry*. Readers would typically not look in the former for compounds with interesting biological activity, nor would they read the latter to find new methods in organic synthesis. Thus, a project that invents new synthetic methods that lead to interesting biological compounds is logically split into two papers. This must be done carefully, however. For certain, a journal such as the *Journal of Medicinal Chemistry* mandates that you provide the synthetic protocol for all compounds whose biological activity is being reported. Thus, the actual synthesis of these compounds must be reported in this journal. But, a paper that describes the work done to develop those methods, along with any other (presumably non-biologically active) compounds being separately reported in a journal such as the *Journal of Organic Chemistry*, is wholly appropriate. This will allow both fields to better find the work. The synthetic crowd can easily find the report in the *Journal of Organic Chemistry* and may be completely uninterested in the biological activity of the compounds if their goals are not biological in nature. The medicinal crowd on the other hand would be unlikely to browse the *Journal of Organic Chemistry* and thus would lose the opportunity to find these important biological compounds if the work wasn't split.

HYPOTHETICAL CASE: SIMULTANEOUS PUBLICATION AGREED TO

A research team has a diverse array of projects, including the invention of new synthetic methodology and the discovery of new compounds with anticancer potential. After years of hard work, they are able to devise a new synthetic methodology that allows them to prepare new compounds with outstanding activity against kidney cancer cells. The leader of the research team separates the two topics, the synthetic methodology and the anticancer properties, into two different manuscripts and submits them to the journal *Science*. They argue in their letter to the editor that the two topics warrant different publications because both are large enough breakthroughs to warrant their own manuscript, otherwise one may get lost at the expense of attracting an audience to the other. Each paper cites the other.

Pondering Points

- What should the editor do?

- Is it appropriate for the two papers to cite each other?

- Does the justification by the authors hold water?

Wrapping Up

For diverse research teams, such a scenario is not impossible and may in fact even be common. The situation described is a classic example of when it would be absolutely appropriate to separate this research into two reports. New synthetic methodology is always important. Such breakthroughs allow for chemical transformations heretofore impossible or so low-yielding as to be worthless. Likewise, drug candidates with outstanding activity or that address an unmet medical need are discovered all the time, and the importance of such a discovery likely doesn't need explanation. Combining the two reports into one paper, though not inappropriate, *may* cause each discovery to be missed by the portion of the scientific community most likely to benefit from it. How? If the report was published in a synthetic chemistry journal, the medicinal chemistry community is unlikely to notice. There is no reason for that community to regularly browse the synthetic literature. Similarly, if the report was published in a medicinal chemistry journal, the synthetic crowd is unlikely to benefit from it. Once again, the synthetic community has no compulsion to regularly check the drug discovery publications for synthetic techniques. Certainly, there are high-profile general science journals such as *Science* that such a research project could be published in. However, journals such as (and specifically) *Science* are *extremely* selective about what they publish. Even with its high impact factor (effectively, a measure of how many times, on average, an article in any issue is cited), a report that details both discoveries at once may not be most easily found by the community it would best serve. Since communicating and subsequently furthering science is one of the most important steps in the scientific enterprise, it is often the better tactic to publish it in a way that will increase its impact on the field(s).

HYPOTHETICAL CASE: AUTHORSHIP ISSUES—AUTHOR OBJECTIONS

Consider the following scenarios:

A research team begins to prepare a report of a team project. As the paper enters into the final draft, one of the co-authors begins to object to the layout of one of the data tables, insisting that an alternate layout is more readable. This opinion is shared with nobody, but the co-author refuses to permit publication on the electronic verification system the journal uses.

A research team is in the final draft stage of a major publication that demonstrates a never before observed phenomenon. After the team began writing the report, their institution acquired a new state-of-the-art instrument that would categorically confirm or refute their discovery. Unfortunately, the instrument will not be available for another month. One co-author insists they wait and refuses to sign off on the publication through the journal's verification system.

Pondering Points

- Which, if any, objections to publication are reasonable?

- What recourse may there be in either case?

Wrapping Up

In cases such as objections to writing style, aesthetics, or just general non-substantive qualities of a paper, objections can be considered unreasonable. All of these are issues do not alter the science or interpretation, even if it *may* impact the readability of a paper. The peer reviewers are within their authority to point out such readability deficiencies if they detect them. Placing an alternately formatted table in supplemental information of a paper can be a viable option. In extreme cases, the PI can appeal to the editor and some sort of addendum can be added to the paper stating that an author objects to the layout of certain data. Even substantive objections can be voiced in this manner, in fact. However, substantive objections should be taken more seriously and can rightfully hold up a publication. Without a really good, *scientific*, reason to not wait, scientific issues that can be resolved prior to publication should be. Sometimes, it may be sufficient to answer those concerns in the future and say you plan to do so. But, 1—you must actually do so, and 2—it needs to be defensible to do so. Avoiding being scooped (someone else reports the discovery before you do) doesn't rise up to the threshold to not wait. Winning priority battles at the risk of lower quality science harms the scientific enterprise.

HYPOTHETICAL CASE: AUTHORSHIP; AN AUTHOR IS NOT CONTACTED

A research team writes up a report of a multi-year project. Over the years, several temporary summer research students work on this project. All the students whose data are being used are added as co-authors, but no drafts are ever communicated to them. The team has a goal of getting the publication out by the end of summer and they are concerned it will take the excluded students too long to respond or track down an active email address. The paper is ultimately accepted for publication. Even at this point, the temporary research students are not all contacted.

Pondering Points

- Can these co-authors benefit professionally from a paper they don't know exists?

- Can these co-authors reasonably be held accountable for errors if they don't know the paper exists?

- Can any complications arise, for the co-authors or the work in question, because of this oversight?

- Assume for a moment that one of the excluded student authors now has a conflict of interest with the work being published, a conflict that did not exist at the time they were performing the work. Can this student be blamed for not declaring the conflict?

- What length of time is reasonable to wait for someone to respond to a draft? Does it depend on the size of the draft?

Wrapping Up

All authors must give permission for a publication to move forward. More and more journals perform this check electronically, but this is not a foolproof system. For example, some nefarious researchers have taken to adding famous researchers and peer reviewers, by faking email addresses to "give approval" or "review positively," respectively. If it is not already happening, it is only a matter of time that this is done in cases like this as well. It is also wholly inappropriate to give someone short notice to review a manuscript; at least two weeks should be allowed to permit someone to comment. A large part of why this checking is necessary is that, as a co-author, you are taking responsibility for the work, not just your contribution. This doesn't mean you need to be an expert on the whole work, though. Also, conflicts of interest can be present either when a paper is being written or when the work is being done, or both. If a potential conflict existed at any of these times, it must be declared in the paper. If a co-author is not aware of a publication, they cannot declare such a conflict of interest.

HYPOTHETICAL CASE: AUTHORSHIP; USING TRICKS TO APPROVAL

An early career professor submits a manuscript to a peer-reviewed journal, and it is rejected by the editor without review. The professor makes some significant additions after performing some additional experiments and, this time, the editor sends it to referees. After review, the paper is rejected on grounds of not significantly adding to the body of knowledge in the field. Frustrated, the professor adds a colleague's name to the author list, a colleague who is well-renowned in the field and has no idea this has happened. The paper is submitted to a higher-profile journal where it is accepted with minor revisions. During the revision process, the professor removes the famous colleague claiming that the revisions actually remove the portions the famous colleague contributed. The as-described manuscript is eventually published in a further issue of the journal.

Pondering Points

- What, if anything, did the editor do wrong?
- What, if anything, did *any* of the peer reviewers do wrong?
- What, if anything, did the professor do wrong?
- What, if anything, did the famous researcher do wrong?
- Does the fact that the professor eventually removed their famous colleague's name make it all OK?

Wrapping Up

Believe it or not, something like this has happened.* Right or wrong, it is generally true that work done by high-caliber, famous researchers garners less criticism than work done by less-established researchers. For sure, some of that is deserved as there is an assumption that their work is always high quality or on the cutting-edge. A reviewer may then feel like *they*, not the famous researcher, must be wrong; or an editor may be highly incentivized to make sure that their journal publishes *the next great discovery*. To a very real extent, this is a type of bias, but neither is automatically unethical, though they can be. For sure, however, adding such a co-author—even if they are removed later—is always unethical, if the person contributed nothing to warrant authorship.

---

* https://retractionwatch.com/2016/11/28/new-way-fake-authorship-submit-promi nent-name-say-mistake/ (last accessed May 5, 2019).

HYPOTHETICAL CASE: AUTHORSHIP; TOO GENEROUS

Consider the following pair of cases:

A research lab has many students and research associates. As a rule, the PI adds every worker in the lab to the author list for all reports in peer-reviewed journals, even if they are working on other projects. Their reasoning is that since everyone works as a team, nobody could do their job without everyone else's help; either instrument upkeep, ordering chemicals, laboratory upkeep, or intellectual contributions.

HYPOTHETICAL CASE: AUTHORSHIP; APPROPRIATE GENEROSITY

Several years ago, an institution hired a new director of a particular instrument facility. Since then, the quality of data the students in one particular group acquire has increased dramatically and they are often trying new experiments suggested by this director. The PI of the research group adds the director as a co-author to all papers that used any of the instruments in the suite they direct.

Pondering Points

- In either case, is a PI overly generous or the tactic reasonable?

- What impact may this have on the perceived contribution of the major (i.e., actual) contributors on the project especially in the former case. In the latter case, will it insinuate an expertise in the field that the director actually lacks as it extends beyond their directorship?

Wrapping Up

Although it is true that the PI has the final say in authorship, there are conventions that most researchers adhere to. Generally, to earn authorship, one must substantively contribute to the publication, be it a paper in a peer-reviewed journal, a seminar, or a poster presentation, the last two may have the presenting author listed only with clear articulations made regarding who contributed what throughout the presentation or at the end. The official proceeding, however, in the case of posters and presentations at official meetings, really should list all the authors to attribute proper credit. What constitutes a significant contribution is at the discretion of the PI, but a wise PI would make their expectations clear very early in the research, certainly before writing has started. This is from both perspectives: Who is listed as an author and who is not. Awarding authorship to undeserving workers acts to reduce the perceived import of every individual's actual contribution to the project. Although it can be argued that attribution is nothing more than hubris, it also puts individuals who have had nothing or little to do with the work into a position where they are responsible for the work; a responsibility they may lack the expertise and/or experience to fulfill. For this reason—accountability—limiting authorship can be very important but shouldn't be overdone to the point of denying deserving authors their credit.

HYPOTHETICAL CASE: AUTHORSHIP; LEAVING OFF A BAD COLLEAGUE

A research group is ready to publish the results of a multi-year study. One of the cornerstone protocols (which is not reported elsewhere yet) is that of a person who left the lab a few years earlier on bad terms. The individual contributed no data, but all the data was collected by other workers using this protocol. To avoid having to work with this individual in writing the paper, they as a group, and with the support of the PI, leave this individual off the author list and simply mention them in the acknowledgment.

Pondering Points

- Was the research team justified to leave this former colleague off the paper as a co-author, especially since they don't work in the lab anymore?

- Is the acknowledgment sufficient?

- Since the PI ultimately makes the decision, is the research team also culpable for this decision?

Wrapping Up

No matter how insufferable a co-worker may be, if they have collected data, created essential and novel protocols, contributed extensively to the writing, or otherwise contributed substantive intellectual ideas, they are deserving of co-authorship. To leave them off a paper to avoid collaborating with them to write is a violation of authorship conventions. One potential gray area is that of novel, essential protocols. A possible solution would be to report such a protocol separately, with the former co-worker and then cite this report in all subsequent work. However, this is a very gray area and an argument can easily be made that this is insufficient; if the protocols have not changed, the individual is still contributing to the research.

HYPOTHETICAL CASE: WEB 2.0 POSTING OF A RESEARCH RESULT

A researcher fed up with how long it took (several months) for their last paper to navigate through the peer-review system decides to post the results of their latest research on their professional blog. Within days, they receive more comments and engage in more discussion about this research than all of their twenty-some-odd previously (and traditionally) published manuscripts combined. Some of the discussions in the comments thread of the blog lead to additional experiments; some of which confirm the conclusions, but a few challenge them, causing the working explanation to be tweaked. One particular thread leads to a new collaboration with researchers in another country.

Pondering Points

- Is there anything wrong with choosing to post your results in this fashion, rather than let peer review vet it first?

- Is the discussion of the results in the comments thread a form of peer review?

- How, if at all, is this different from a press release made through the popular news media?

Wrapping Up

The use of Web 2.0 (blogs especially) is becoming an increasingly considered mode of scientific communication. Some journals have taken to using an open review of manuscripts, which takes advantage of the same qualities that make Web 2.0 such an attractive venue for publications. First, in Web 2.0, publication is instantaneous. There is no need to wait for peer review. While this may be considered a negative from the point of view of possibly allowing junk research to enter a publication record since these reports are not reviewed before they are made public, it should be noted that the comments that are made effectively act as a kind of peer-review system in a way that ordinary peer review does not. In this Web 2.0 type approach, the reviews are all public and can be responded to—not just by the author, but by the other readers as well. This allows for a public and potentially international discussion to occur about the research. This is far more discussion than can typically happen with a traditional publication. Citing blogs is as easy as citing any other references, so it shouldn't impact any other researchers who try to further or otherwise reference the work in question. Other than an attachment to traditional publications, fueled by either hubris or an unwillingness to change, there is no reason to not pursue such a publication tactic. One can certainly argue that the number of times a research report is cited is a measure of its worth and that this may be more difficult in the blogosphere. However, as mentioned before, blogs can be cited, and it is also easy to argue that a research report posted as a blog that generates thousands of comments is at least as impactful as any traditional research report cited by twenty researchers.

HYPOTHETICAL CASE: COPYRIGHT ISSUES CREATIVE COMMONS

A professor writes a book that is electronically published on their website and registered with the copyright office. They then offer access (by way of a free account on their website) to the book to anyone who is going to use it for educational purposes, provided that they do not copy it onto their own website and also that they do not monetarily gain from the work. The professor who authored the work rents out advertisement space on their website, earning money off their work this way on a per-ad click basis. They also embed their own YouTube videos on their website, which earns them additional profit on a per-view basis.

Pondering Points

- Are the terms of the license unreasonable?

- Is it inappropriate for the author to earn money in this way off their product?

- Is it inappropriate to not use an official publisher?

Wrapping Up

This is actually referred to as a Creative Commons license, though the self-run website is not a mandatory component. There are generally four basic restrictions that are used in Creative Commons licensing. The first is Attribution—you must give the author credit. The second is Non-Commercial—that is, you cannot profit off of someone else's work. The third is No Derivatives—this keeps someone from modifying the work; modifying the work makes something open source. Finally, the restriction of Share-Alike—which allows users to modify the work, provided they let others share and further modify the work, which is a bona fide open-source approach. These last two are not in conflict with each other since generally only Attribution is always part of the license. The others can be mixed and matched as the author desires, save for mixing Share-Alike and No Derivatives, which are opposites. The author in our story here gains the attribution by mandating that all the users go through their website. Profiting from your own website's traffic and/or YouTube videos is in no way unethical. In fact, as resources additionally move to a digital medium, it may become more common.

HYPOTHETICAL CASE: COPYRIGHT ISSUES FAIR USE

A professor is teaching a special topics class that has three students enrolled in it. As part of the class, they are analyzing a set of short stories by one particular author. All of these stories, along with a few others, are published in a best-selling anthology. The set of stories the class will analyze comprise a little more than half of the anthology. The professor copies all the stories the class will analyze, along with the copyright page from their personal copy of the anthology, and distributes it to the students on the first day. The students then use these copies throughout the semester. On the teaching evaluations at the end of the semester, the students comment how nice it was to not have to buy the whole book as it helped them save some money.

Pondering Points

- Is this plagiarism?
- Is this theft of material?

Wrapping Up

Under the conditions described here, this would almost certainly fall into the category of fair use. The courts would ultimately decide, but for several reasons, according to the U.S. Copyright Office, this falls into an area that is not considered theft of material or plagiarism. To evaluate a use for application to the fair use clause, courts look at several factors. One is the purpose and character of the use, including whether the use is of a commercial nature or is for nonprofit educational purposes. In our story here, this clause is easily applied. Another is the nature of the copyrighted work. Unpublished works or creative works like a novel or movie are less likely to be supported than some manner of technical work. On some level, this may count as a strike against our story in that it was short stories that were copied. Third is the amount and substantiality of the portion used in relation to the copyrighted work as a whole. Court history has no "go-to" amount that is OK. According to the U.S. Copyright Office,* the courts have found the use of an entire work to be fair and the use of a small portion to be unfair. Finally, the effect of the use upon the potential market for, or value of, the copyrighted work. Here, it is likely the ace that wins this use as fair use in this case. That the class had three students in it hardly will impact the market share and value of a best-selling work. Nevertheless, the court may surprise. The fact of the matter remains that, educational uses, especially in small classes, are likely to be regarded as fair use. However, it would be wise to always check.

HYPOTHETICAL CASE: PEER REVIEW BAD REVIEWER

A peer reviewer is reviewing a paper that is very closely related to ongoing work in their own laboratories. They delay submitting the review of the paper until their research group completes their own work and submits a paper of their own. They also recommend that the authors of the paper they reviewed complete some additional studies and include more information in the introduction, including papers their own lab has previously published on semi-related work.

Pondering Points

- Was the delay of the review appropriate?

- Should the reviewer have let the editor know about their lab concurrently working on such a similar project?

---

* https://www.copyright.gov/fair-use/more-info.html (last accessed May 5, 2019).

- Was the suggestion to add more experiments appropriate?

- Was the suggestion to add more references to the introduction appropriate?

Wrapping Up

This is an absolute abuse of the peer-review system. Reviewers are expected to be timely with their reviews and declare if there is a potential conflict of interest. While for some fields (especially small ones), some level of conflict of interest is inevitable, the editors must always be made aware of any, even potential, conflicts. Reviewing a competitor's work certainly fails into the realm of conflict of interest. After the potential conflict is declared, it is up to the editor regarding how to proceed. A conscientious and good editor will be on the lookout for inappropriate behavior and pursue other reviewers if such behavior is observed. Recommending additional work or references in an introduction is not inherently inappropriate. Doing so in a way that delays the publication to help yourself but doesn't necessarily add to the publication in question, on the other hand, is. In the case of adding to the introduction, there are circumstances where recommending your own work be cited is appropriate, but these are somewhat limited. The work must be directly relevant to the manuscript being considered. An editor can of course override the suggestion, however.

HYPOTHETICAL CASE: FAKE PEER REVIEW

A PI is worried that a manuscript they are about to submit will be reviewed harshly. On the list of recommended reviewers, they invent a fictional person and add them to the list, including an email address they still have access to from a previous employment. They then log in as this fictional reviewer and give their own paper a positive evaluation.

Pondering Points

- Why is this inappropriate?

Wrapping Up

Perhaps this should be obvious, but you cannot pose as a reviewer and review your own work. There is nothing appropriate or defensible about such behavior, particularly since, in order to do so, you must either wholesale invent a person or masquerade as a real person. Since it is additionally inappropriate to serve as a reviewer for a colleague, or some other co-worker or closely associated person (think family member, etc.) as it could be a conflict of interest, this is essentially the most extreme version of a conflict of interest. This is because you're now the reviewer (the official evaluator) of your own work.

Index

Printed in the United States
by Baker & Taylor Publisher Services